Stécie Audrey MALIEZE

Modeling and simulation of batteries in a photovoltaic system

Stécie Audrey MALIEZE

Modeling and simulation of batteries in a photovoltaic system

ScienciaScripts

Imprint
Any brand names and product names mentioned in this book are subject to trademark, brand or patent protection and are trademarks or registered trademarks of their respective holders. The use of brand names, product names, common names, trade names, product descriptions etc. even without a particular marking in this work is in no way to be construed to mean that such names may be regarded as unrestricted in respect of trademark and brand protection legislation and could thus be used by anyone.

Cover image: www.ingimage.com

This book is a translation from the original published under ISBN 978-620-3-43232-9.

Publisher:
Sciencia Scripts
is a trademark of
Dodo Books Indian Ocean Ltd. and OmniScriptum S.R.L Publishing group
Str. Armeneasca 28/1, office 1, Chisinau MD-2012, Republic of Moldova, Europe
Printed at: see last page
ISBN: 978-620-5-38710-8

Contents

Dedication

MESPARENTS

Acknowledgements

First of all, I would like to express my gratitude to God Almighty for his blessings and all the achievements he has bestowed upon me.

I would like to thank in particular my supervisor, Prof. TCHINDA Rene, who guided me with professionalism and efficiency during my research work.

My thanks are also addressed to Dr. ALOYEM KAZE Claude Vidal, for his assistance in the work, for his availability, his dedication and for his precious advice. Thank you very much.

I sincerely thank the members of the jury who sacrificed their time to judge the quality of this Memoir.

I would also like to thank Mr. FOGNO FOTSO Hervice Romeo for his assistance and help in the completion of this work.

Thank you to all my fellow Master students for the high quality work atmosphere to which they have contributed with enthusiasm and whose results are perceptible at this very moment.

I also thank my mother DJAMEN Florentine, my father DJIAMBOU Jean Claude and my aunt DJOUKO Sylvie for their encouragement and support. Thank you to MELI Noel Rivarol, TOUOTSAP Arsene, a papa TALELEMOULA Gaspard, maman JIPAP Angele and my brothers and sisters for their love and encouragement : ASSOGO Ange, KENDA Gides, SIANI Bruce, WANJI Naomie, NGUEYI Sherifa.

Resume

The demand for electrical energy is growing worldwide and in Cameroon in particular. Fossil energies being exhaustible and harmful for the environment, clean energies such as photovoltaic energy constitute a solution of relay by guaranteeing an unlimited and non polluting reserve. As photovoltaic energy is intermittent, it is necessary to design a storage system, the most common of which is the accumulator battery. The most common storage system is the accumulator battery. This half of the system needs to be controlled and managed permanently for efficient use; hence the need to model it in order to highlight all the phenomena within it. In this work, we present three electrical models for the moderation of the lead acid battery in Eseka, Cameroon, allowing us to study its operation and evaluate its performance. As the battery may suffer from premature degradation, we estimated its state of charge as well as its charging and discharging voltages. As there is no single model that brings out all the phenomena of the battery, we used the CIEMAT model to study the influence of exogenous parameters such as temperature and current intensity on the battery; the Thevenin dynamic linear model allowed us to study the temporal evolution of the voltage and the Thevenin improved dynamic model allowed us to study the self-discharge phenomenon within the battery. On the basis of the simulation results obtained, it appears that even an unused battery discharges. In addition, variations in temperature and current have a considerable influence on the performance of the battery. For the Eseka site, the variation of the ambient air temperature does not have a great influence on the performance of the lead acid battery studied for this site.

Keywords: Photovoltaic system, Accumulator batteries, Modelling, Electrical models, Performance.

General introduction

The demand for energy is growing in proportion to population and industrialisation worldwide, and in Cameroon in particular. In addition, the most commonly used energy sources, such as fossil fuels and nuclear energy, are expensive and have negative impacts on the environment, such as pollution, global warming, etc. The main source of commercial energy in Cameroon is hydro (80%). This is insufficient and requires considerable investments for the transport of the production plants to the consumers. Thus, to achieve its emergence by 2035, while preserving the environment, Cameroon needs to develop other sources of energy. Photovoltaic energy is a promising alternative to other forms of energy, which are considered polluting and non-renewable. The development of photovoltaic energy is following a considerable growth. It is used for water pumping, rural electrification, vaccine refrigeration, etc. The autonomous photovoltaic system is characterised by the fact that the energy is produced solely by the photovoltaic panels. Therefore, a storage system is needed to store the captured energy to ensure power supply during the night or in periods of low sunlight.

The storage of electrical energy is a major challenge, and there are several types of batteries commonly used in PV systems. Thus, batteries capable of having a self-contained energy reserve are the most commonly used batteries for energy storage in PV systems [1], in these systems, these batteries operate under the charge/discharge principle. Their performance can be affected by several electrical and chemical factors. Thus, too much charging or deep discharging can lead to premature ageing of the battery [2], hence the need to look for models to represent these in order to ensure efficient management and control of them. However, the difficulty in moderating a battery lies in the nature of the electrochemical and/or dynamic phenomena that occur during its operation. Indeed, to understand the behaviour of a battery, it is necessary to build a model capable of predicting and simulating its operation. Thus, there is a wide variety of battery models that can be used to monitor its state in a PV system [3],

Several battery models are presented in the literature to model them. Thus, Rami

Haddad et al. 2015,[4] have modelled the lead acid battery by a new flexible model in order to study and predict its behaviour through neural networks. Similarly, H.L Chan et al. 2000,[5] presented a dynamic model of the battery offering a strategic approach to reduce large charging frequencies and discharge currents. M.E Glavin et al. 2008,[6] studied the hybrid battery-supercapacitor storage system in order to optimise the system, reduce the battery fault and increase the battery lifetime with a proposed energy control unit. From these studies, it appears that the performance of these batteries is also influenced by the (meteorological) parameters of the operating site, such as temperature. Thus, it is necessary to model the battery taking into account the parameters of the operating site.

Our work consists of developing some electrical models for the study of batteries intended for use in stand-alone PV systems at the Eseka site, coastal region, Cameroon. The internal parameters of the batteries were identified in order to highlight the phenomena that contribute to their degradation. We have implemented the Thevenin's improved model, the Thevenin's linear dynamic model and the CIEMAT model in the Matlab environment. To study the battery we not only considered its state of charge but also its voltage during charge and discharge. We used a battery with a capacity of 160 Ah. The behaviour of the battery is shown by varying some parameters and studying their influence on this half-value.

The present work is divided into three chapters.

The first chapter presents the generalities of the photovoltaic system and the different storage modes. Emphasis is placed on battery storage and some battery types used in PV systems are presented.

The second chapter presents a study of the electrical models and the modelling of the batteries. Thus, three electrical models of batteries, namely the CIEMAT model, the Thevenin dynamic linear model and the Thevenin improved model, were studied and the mathematical equations obtained made it possible to highlight the characteristics of these.

The third chapter is devoted to the simulation results. In this part, we present a revolution of the battery parameters and the impact of the characteristics of the Eseka site such as temperature on its performance for each of the three models studied.

Chapter I

General information on photovoltaic systems and storage modes

Introduction

Nowadays, there are still households without access to electricity and areas without sufficient supply. However, there are many natural resources that can solve the problem, in particular the solar resource that allows the decentralisation of production plants.

Discovered in 1839 by the French physicist BECQUEREL, the photovoltaic effect allows the direct conversion of the light energy of the sun's rays (photon) into electricity (volt), through the production and transport in a semiconductor material of positive and negative electric charges under the effect of light. Figure 1.1 shows the working principle of the photovoltaic effect. In fact, this material has two parts, one with an excess of electrons and the other with a deficit of electrons, known as n-type and p-type respectively. When the former is brought into contact with the latter, the excess electrons in the n-type material diffuse into the p-type material. The initially n-type area becomes positively charged, and the initially p-type area becomes negatively charged. an electric field is thus created between them, which tends to push the electrons into the n-type area and the holes into the p-type area, and if the sides of the junction are connected to a load, a current f flows through it and a potential difference arises [7]. Therefore, the need for electrical energy is solved by the use of photovoltaic systems to provide electrical energy while using solar energy.

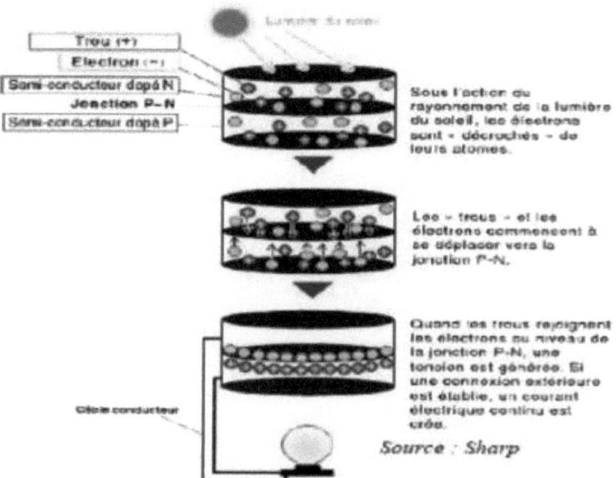

Figure 1.1: The photovoltaic effect[7J

I- The photovoltaic system

The photovoltaic system is a solar installation designed to convert solar energy into electrical energy. Its principle is based on the adaptation of the energy supplied by the panel to be useful for the supply of electrical appliances and even for injection into the electricity grid.

- -1) Elements of a photovoltaic system

A photovoltaic installation as shown in figure 1.2 is generally composed of :

- A photovoltaic generator or solar panels;
- An electrical energy storage system ;
- A charge and discharge regulator;
- An inverter or DC/AC converter;

- Charges to be fed.

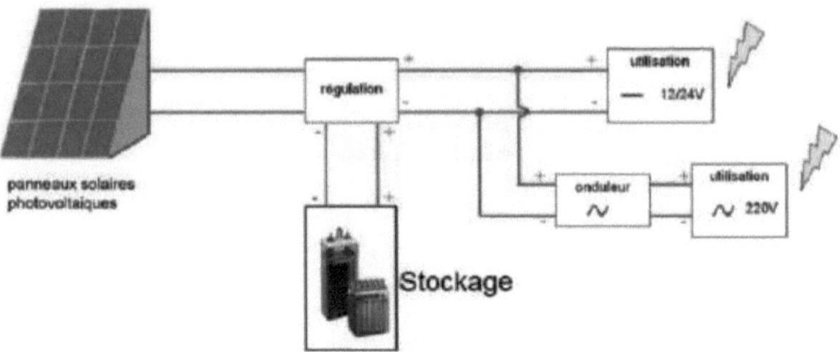

Figure 1.2: Simplification of the PV system[8].

- The PV panel

PV panels or modules are made of photovoltaic cells of a semi-conductive material that transforms the radiation from the sun into DC electricity. The panel is covered with a transparent material that protects it from water and is designed for easy assembly[9]. The efficiency of a PV panel (PV module) depends on several factors [11]:

- Temperature: The effect of temperature on the panel varies according to illuminance, ambient temperature, wind speed, module mounting (rooftop or aerial) and all of these parameters change depending on the site chosen for the installation of the modules. It can be seen from figure 1.3 that an increase in temperature causes an apparent decrease in open circuit voltage.

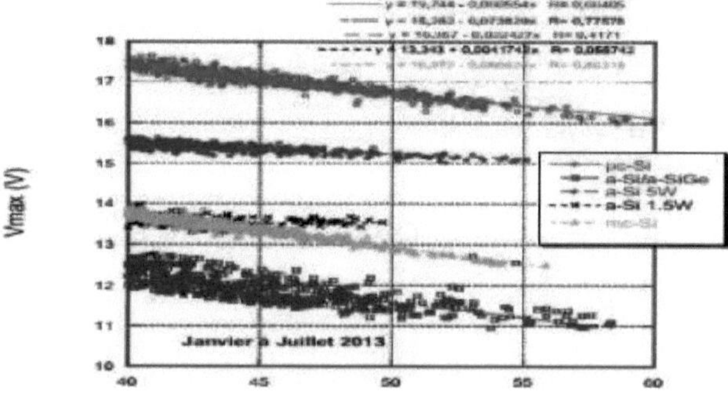

Temperature Panels (°C)

Figure 1.3: Maximum voltage of 5 PV modules as a function of cell operating temperature

- The angle of incidence of the radiation: Whatever the module, the incident light must pass through several successive layers before it can reach the cell, so the light rays can come from anywhere, which generates a loss proportional to the cosine of the angle of incidence, but also a loss due to reflections on the glass which can be added to the overall losses for angles of incidence greater than 50°.
- Partial shading: shadows and dirt from the site of installation reduce the performance of the PV module

❖ The inverter

An inverter is a power electronics device that provides alternating voltages and currents from an electrical energy source of different voltage or frequency. It is the reverse function of a rectifier. The inverter is a static DC/AC converter as shown in figure 1.4 [8].

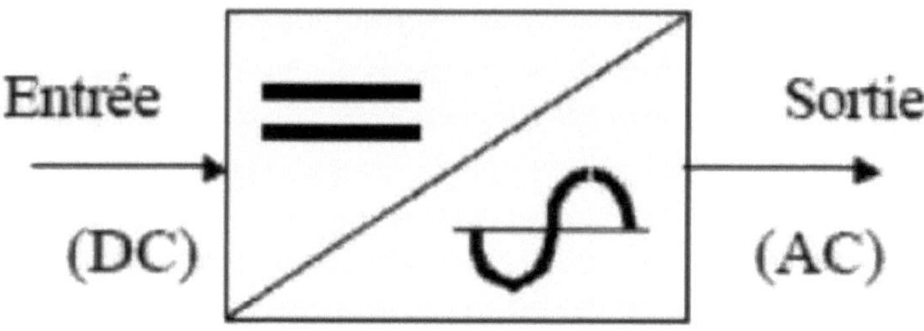

Figure 1.4: Symbolic diagram of the inverter[8J

❖ Charge and discharge regulator

The charge controller protects the system's investment and has a direct impact on the life of the batteries. It regulates the energy from the panels to the battery, stops charging the battery when it is fully charged and cuts off the energy from the battery when it is below the safe level. A robust energy charge controller with a predefined feedback can optimise the energy required to be distributed by the PV system and thus make it available [9].

❖ The storage system

It is of vital importance in a photovoltaic installation. It is used to supply loads during periods of low sunlight, at night or during an interruption in the electricity supply.

1-2) Types of photovoltaic systems

Photovoltaic systems are used today in various applications:

- Homes: Photovoltaic systems are an economical option for isolated homes where the available infrastructure does not allow access to electricity;
- Mobile and recreational applications;
- In agriculture: effective in pumping water for livestock, plants, or humans on hot sunny days.
- Photovoltaic systems can also be used for other purposes. Photovoltaic cells can be found in watches, calculators, power banks, road construction panels, parking lights etc.[12].

There are three types of photovoltaic systems: stand-alone or isolated, grid-connected and hybrid.

> An insulated photovoltaic system

Shown in Figure 1.5, the stand-alone photovoltaic system supplies the user with electricity without being connected to the grid. It is often the only way to get electricity when grid power is not available: houses in remote locations, on dunes, in the mountains. This type of system requires the use of batteries to store the electricity and a charge controller to ensure the durability of the batteries [8].

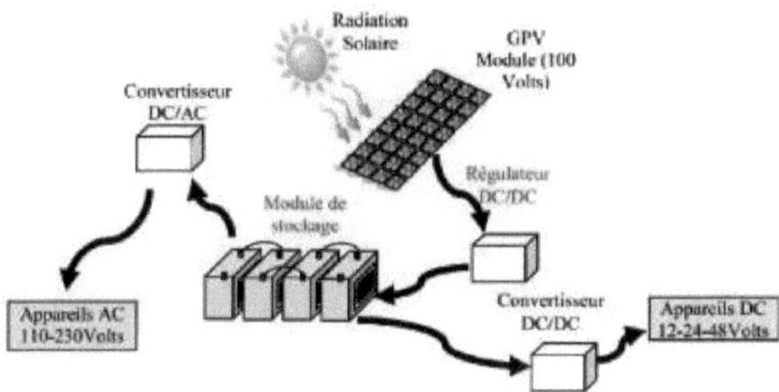

Figure 1.5: Isolated photovoltaic system [13]

> A grid-connected photovoltaic system

It is a system coupled directly to the grid by means of an inverter and is shown in Figure 1.6. This type of system offers a lot of convenience for the producer/consumer as the grid is responsible for balancing the production and consumption of electricity. In the case of grid-connected systems, it is imperative to convert the direct current produced by the photovoltaic system into an alternating current synchronised with the grid. To perform this conversion, an inverter is used [8].

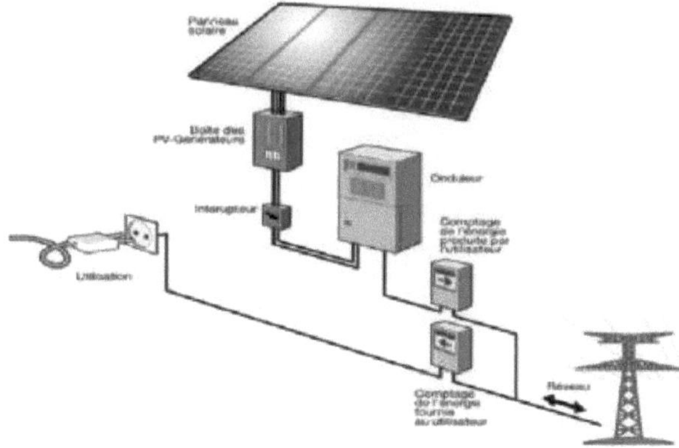

Figure 1.6: Grid-connected photovoltaic system[8]

> The hybrid photovoltaic system

The hybrid energy production system, in its most general view, is one that combines and exploits several readily available sources. It consists of the combination of two or more complementary technologies in order to increase the energy supply through better availability. Energy sources such as solar and wind do not deliver constant power, and their combination can lead to more continuous electricity production [7]. The hybrid system often (but not necessarily) includes a storage unit, and is connected to a local distribution network (mini-grid) as shown in Figure 1.7. Since PV panels produce direct current (DC) and mini-grids operate on alternating current (AC), the core of a hybrid system is a multi-functional inverter capable of converting DC and AC currents, controlling the generation and storage systems, and setting the voltage and frequency of the mini-grid [12].

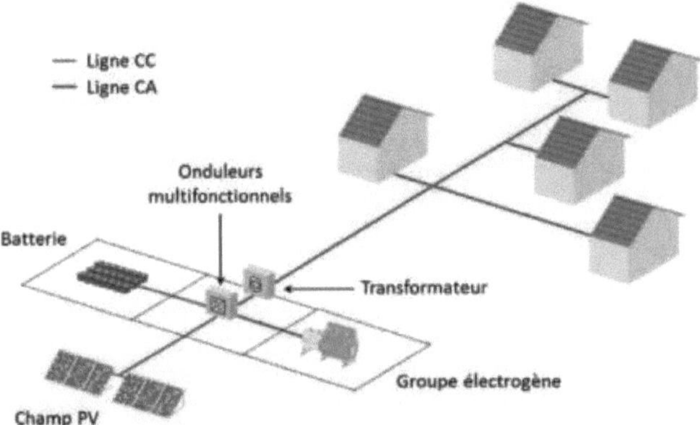

Figure 1.7: PV-diesel hybrid system for rural electrification

1-3) Type of photovoltaic system used in Cameroon

In Cameroon, stand-alone photovoltaic systems are found only in power plants used for public lighting, for the supply of electricity in some enterprises and for surveillance systems as illustrated in Figure 1.8.

Figure 1.8: PV installation for surveillance camera at the central post office, Yaounde

II- The storage system

II-l) The different storage technologies

In a photovoltaic installation, storage consists of keeping the energy produced by the photovoltaic generator on standby for later use. The management of solar energy requires the consideration of storage according to the weather conditions. The fundamental characteristics that condition the choice of a storage system are [10]: Power density (W/Kg), energy density (Wh/Kg), number of operating cycles, cost and energy efficiency. There are several storage techniques:

- Magnetic energy storage in superconducting coils[14]

Superconductors have the property of exhibiting zero resistivity when cooled below a critical temperature Tc. Thus, if a superconducting coil is energised and then short-circuited to itself, the current is not dissipated by Joule effect, and the magnetic energy is conserved almost indefinitely. The stored energy is restored when the coil is short-circuited to a load. This is the principle of superconducting inductive storage, commonly called SMES (Superconducting Magnetic Energy Storage). The stored energy E_{mag} can be expressed as a function of the inductance L and the current I according to the equation :

$E = - LI^{mag}$

It should be noted that the development of superconductors is still limited, which makes them very expensive[10],

- Double layer capacitor storage: super capacitors

The basic principle of electric double layer super capacitors is based on the capacitive properties of the interface between a solid electronic conductor and a liquid ion conductor. Energy storage is achieved by the distribution of electrolyte ions in the vicinity of the surface of each electrode, under the electrostatic influence of the applied voltage. This creates a space charge zone at the interfaces, called an electric double layer, with a thickness of only a few nanometres [15].

- Storage in the form of kinetic energy: flywheel

The flywheel energy storage (EES) shown in Figure 1.9 consists of a solid or composite flywheel (faster, higher mass energy) combined with a motor-generator and special (often magnetic) bearings, all in a very low-pressure containment to minimise self-charging losses. Self-charging rates of a few percent per hour can be achieved. In a simplistic way, the flywheel is sized in energy and the motor-generator in power; energy and power are thus easily "decoupled". The flywheel therefore stores energy in the form of kinetic rotational energy, in a speed range between its maximum speed (about 35,000 rpm) and about half that (15,000 rpm). In order to achieve acceptable energy densities, it is necessary to use extremely strong composite materials to increase the speed of rotation [16]. Flywheel storage is classified as a low time constant system [17].

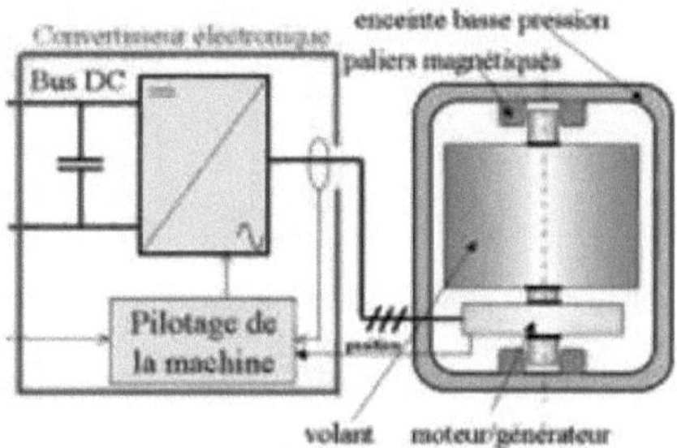

Figure 1.9: Flywheel Accumulator[18]

- Compressed air storage (pressure)

The energy produced by a photovoltaic power plant can be fed into the grid and used to compress (by means of a turbocharger) the ambient air to a high pressure (100 to 300 bar) during periods of low consumption. Once compressed, the air is used to power a turbine to generate electricity [19][20]. The operating principle of the compressed air storage system is illustrated in Figure 1.10.

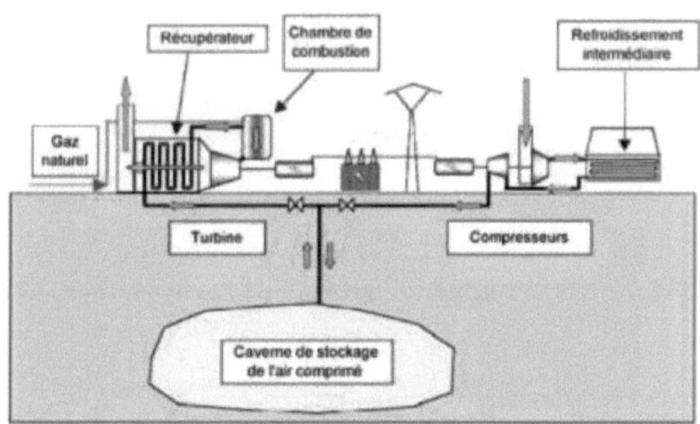

Figure 1.10: Operating principle of a compressed air storage system[20].

- Thermal storage

Also known as thermal pumped storage, this technology uses a heat pump to massively store electrical energy in thermal form in low-cost solid materials. During the charging phase, the electricity compresses a working fluid, the heat is transferred into a split container to provide a large internal exchange surface area relative to the power required. During discharge, the heat is transferred to a turbogenerator [21].

- Storage in gravity or hydraulic form

Pumped Storage Power Stations (PSTS) consist of two water reservoirs at different heights connected by a pipe system. They are equipped with a pumping system to transfer water from the lower basin to the upper basin. The schematic diagram is shown in Figure 1.11. During periods of high sunlight, when the PV field produces surplus electricity, water from the lower basin is pumped to the upper basin. Under the effect of gravity, this body of water represents future electricity generation capacity. When there is a shortage of electricity production in the network, part of the upper reservoir is emptied and, by gravity, the water turns a hydraulic turbine which feeds an altemator and produces electricity.

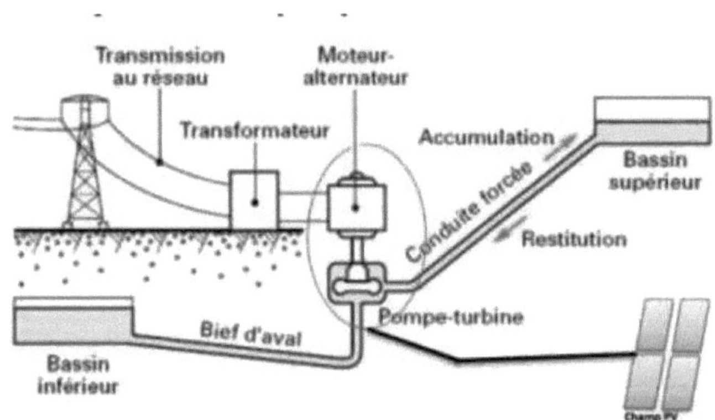

Figure 1.11: Principle diagram of a gravity storage system[20].

- Electrochemical storage: batteries

This is the most common form of storage today. There are different types of batteries, so different chemical reactions are caused by using electrical energy and storing it in chemical form, and then by reverse chemical reactions, electricity is produced. Batteries are an assembly of cells composed of different types of chemical elements. A cell consists of three basic elements: the positive electrode (cathode), the negative electrode (anode) and the electrolyte. Within the battery, ions move between the electrodes through the electrolyte. The purpose of a cell is to create a potential difference between the anode and cathode [10][19]. Table 1 shows the different storage technologies and their advantages and disadvantages.

Table 1: Comparison of different storage technologies[10].

Technologies	Benefits	Disadvantages
Battery	Low cost	Low life span
Compressed air	Large storage capacity	Specific site Flows with natural gas
Hydraulic pumping	Large storage Low cost	Construction time Favourable site
Flywheel	High power	Low energy density Cost
Superconductor	High power	High cost Low energy density
Super capacitor	High life expectancy Good performance	Low energy density

II-2) Choice of storage system

Electrochemical storage has advantages over other systems. In fact, electrochemical storage is easy to adapt to the power and energy needs of the installations; it allows the storage of large quantities of energy per unit of mass. 22] Batteries appear to be a solution for storing energy that can be easily recuperated in an electric form by a provoked phase of discharge, and then regenerated during a phase of charge. Another advantage is that they are recyclable, so that some components can be used at the end of their life to make new ones [13].

Moreover, the technology is well adapted to the Cameroonian environment compared to others, in that the use of batteries as accumulators does not require a particular site and the cost is relatively accessible to the population. It should be noted that energy storage in PV systems is done by means of batteries in Cameroon.

III- Accumulator batteries

A battery is an electrochemical system capable of returning stored chemical energy in electrical form. In addition, the internal reactions are reversible [17]. There are several types of storage batteries used in photovoltaic systems; figure 1.12 shows a photovoltaic battery. Solar batteries store the energy produced by the photovoltaic panels in order to ensure power supply in all circumstances; their discharge is slow. It is possible to connect a solar battery directly to a solar panel, but this may damage the battery if its charge level exceeds 90%. It is therefore strongly recommended to install a solar regulator between the solar panel and the solar battery(ies).

Figure 1.12: Photovoltaic battery

III-l) Electrochemical principle

It is necessary to understand the basic chemical phenomena governing the operation of batteries in order to be able to justify the choice of methods and models, and subsequently better interpret the results of these interactions.

When the battery is discharged, the oxidation reaction that occurs at the anode releases one or more electrons into the external circuit. These electrons then flow to the cathode where they participate in the reduction reaction (gain of one or more electrons). At the same time, anions and cations migrate into the electrolyte solution between the two electrodes in order to maintain the charge balance. When the anode is completely oxidised (or the cathode completely reduced), these reactions are completed and the battery is discharged. To charge the battery, an electric current is applied to the electrodes in order to generate the opposite reactions.

In discharge, the anode is the negative terminal of the battery and the cathode is the positive terminal. On the other hand, in the charging phase, the negative electrode is the cathode and the positive is the anode, with the electrons flowing in the other direction [13]. The schematic diagram of an electrochemical cell is shown in figure 1.13 :

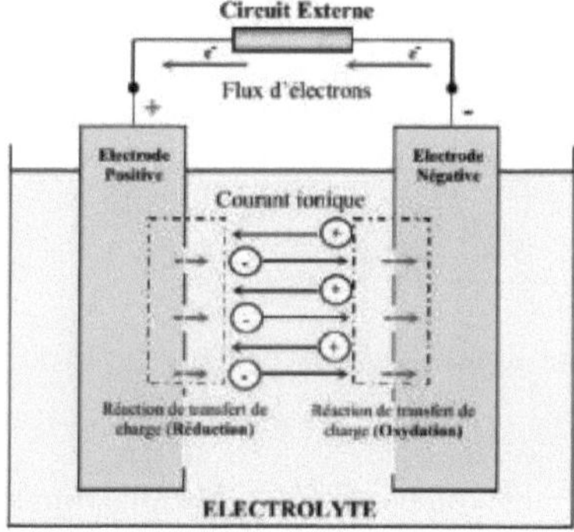

Figure 1.13: Basic electrochemical cell [13].

III-2) Types of battery

III-2-1) Lead batteries

A lead-acid battery is made up of several cells connected in series. The voltage of a lead-acid battery is always a multiple of about 2 volts. Because of its age and wide distribution, it acts as a benchmark for assessing the characteristics of other batteries.

Shown in Figure 1.14, this system consists of two electrodes (positive and negative) and an electrolyte. The positive electrode is made of lead dioxide (PbO2) and the negative one of lead. The electrolyte is a solution of sulphuric acid (H2SO4) which allows the flow of ions between the two electrodes and creates a current. The potential difference between the two electrodes is2V [17].

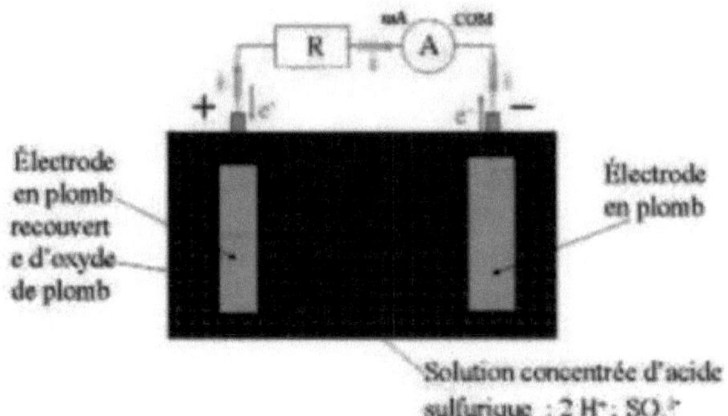

Figure 1.14: Schematic of a lead-acid battery[23].

The electrode-electrolyte assembly is the site of a redox reaction whose condensed global equation is written as follows:

$$PbO_2 + Pb + 2H_2SO, \underline{<\ '\ ^{\kappa e} >} 2H_2O + 2PbSO,^{z\ Z4} \qquad {}_c h_{ar} g_e z$$

Lead-acid batteries are divided into two main families: Vented Batteries and Valve Regulated Lead Acid Batteries [17]. Lead-acid batteries are among the most widely used battery types in photovoltaic systems. The diagram showing the world market for lead-acid batteries is shown in figure 1.15.

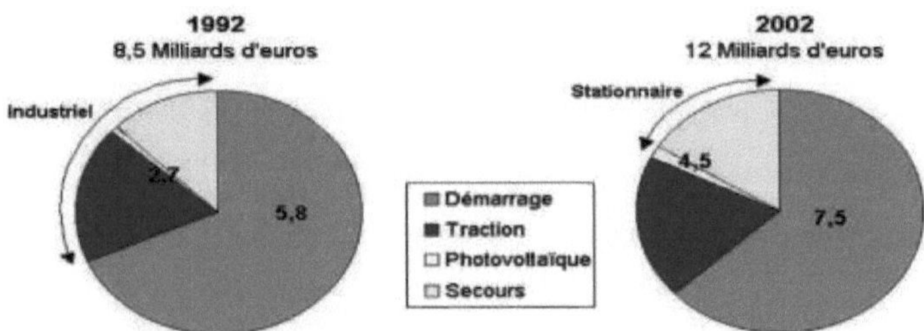

Figure 1.15: World lead-acid battery market[24].

III-2-2) Nickel-Cadmium (Ni-Cd) batteries

Nickel-Cadmium batteries have a physical structure similar to that of lead-acid. Instead of lead, they use nickel hydroxide for the positive plates and cadmium oxide for the negative plates. The electrolyte is Potassium hydroxide. In spite of a still prohibitive price, this type of battery has many advantages [24]: > Very good mechanical resistance.

> Ability to withstand deep discharges.

> No toxic emissions from the electrolyte.

III-2-3) Nickel metal hydride (Ni-MH) batteries

Ni-MH batteries differ from Ni-Cd batteries in that the negative electrode is based on hydrogen absorbed into a metal, and this technology offers interesting improvements over Ni-Cd. Firstly, the specific energy is higher. Secondly, the absence of cadmium makes the reprocessing of the battery at the end of its life much easier and a possible rupture of the cell less problematic. Unfortunately, the cost of this type of battery is still very high. These batteries have been developed with the objectives of increasing capacity per unit volume, favouring fast charging and eliminating environmentally toxic cadmium [10]. Apart from the fact that they are more compact, these batteries are not better than Ni-Cd batteries for solar applications. They are even worse in terms of cold resistance.

III-2-4) Lithium (Li) batteries

The increasing use and diversity of electrical applications has led to the development of new electrochemical storage technologies, such as redox systems, hydrogen storage systems and lithium batteries. Used as an active material at the anode, it allows to obtain batteries with high energy potential. However, its reactivity with the environment (especially with air) makes it a difficult material to handle in its metallic state. There are three main families of lithium batteries: Lithium metal, Lithium ion and Lithium polymer [24]. Table 2 gives the advantages and disadvantages of the different types of batteries.

Table 2: The different types of battery [25].

Type of battery	Benefits	Disadvantages
Lead (Pb)	- Low prices - Solid - Capable of supplying high currents - Standard elements found anywhere in the world trade - Easy to implement rnuvre - Without memory effect (i.e., the recharge whenever you want, at any level of decharge) - Flexibility of use - Excellent price/lifetime ratio (3/4 years) - Don't pollute so well recycle	- Energy density - Weight - Self-charging (approx. 1% per day) - Sensitive to negative temperatures (loss of autonomy up to - 25% at 10°C) - Risk of crystallization of Pb if left unloaded for too long and irreversible loss of capacity

Nickel Cadmium (Ni - Cd)	- Suitable for high load and discharge currents due to their low internal resistances. - Low cost - Mechanical and electrical strength - Easy charging and high tolerance to overloads	- Memory effect - Average energy density - Recycling is complicated by the cadmium which is a heavy metal and pollutant
Nickel Metal Hydride (Nimh)	- Good energy density - High current capability due to low internal resistance (Ni-Cd still has an advantage in this area) - Easy to store and transport - Don't pollute so well recycle	- Fragile as they do not support overcharging, requiring the use of powerful and expensive automatic chargers - Difficult to detect the end of charge - Low life span - High self-charging - Technologic outperformed

Lithium (Li)	- Very high energy density due to the physical properties of lithium - Very low self-charge (5% per month) - No memory effect - Weight - How to use it - Accepts a large number of cycles (up to 1500 for the best) - Low internal resistance	- Very high price - Requires a protection circuit (B.M.S.) to manage the load and unloading atm to avoid the destruction of the elements... costly - Wear even in case of not use

IV- Moderation of accumulator batteries

There are several types of battery moderations in the literature.

IV-1) Chemical modelling

This model takes into account the electrochemical phenomena involved in the battery such as diffusion, polarisation and mass transfer within the electrochemical couple.

IV-2) Empirical modelling

Empirical modelling is a classical method based on experimental tests. The performance of the battery is recorded and tabulated. This type of modelling does not represent a generic modelling for all batteries as it does not take into account all parameters, it has to be tested for each type of battery [26].

IV-3) Electrical modelling

Electrical modelling uses an equivalent circuit to represent the dynamics of the battery. This type of modelling is well known and widely used depending on the application and type of battery. Several models have been developed by scientists and each of them generalises its model for other types of batteries.

V- The parameters of a battery

Determining the performance of a battery requires several criteria. In fact, as batteries are used in many fields, it is essential to go through these parameters to classify them according to their type.

> **Nominate voltage:** This is the reference voltage of the battery.

> **Open circuit voltage**: This is the voltage at the battery terminals without any load applied. It depends on the state of charge of the battery.

> **Internal resistance:** This is the resistance inside the battery. It depends on the charge and discharge, but also on the state of charge of the battery. Indeed, when the internal resistance increases, the efficiency of the battery decreases [27].

> **Capacity:** This is the amount of electricity that a fully charged battery can circulate during a given discharge period up to a cut-off voltage at a defined temperature [28]. It is the product of the discharge current and the time until the end-of-discharge voltage is reached. It is expressed in Ah. There is also the nominate capacity which is given by the manufacturer, it corresponds to a specific condition of use.

> **The state of charge:** This is the capacity present in the battery. This half of the battery is considered as a reservoir of energy whose quantity is constantly changing [13], > **The self-discharge phenomenon:** This is the loss of capacity by the battery over a period of time without express intervention [29]. It is interpreted as a loss of

electromotive force due to an internal leakage of current [13],

> **Temperature:** Temperature has an influence on the nature of any element. In the case of batteries, it influences their behaviour and characteristics; the specific temperature is about 27° C; very high temperatures can lead to an increase in water loss and thus decrease the life of the battery, while very low temperatures can freeze the battery causing it to decrease or stop functioning [8].

> **The polarisation phenomenon:** it is expressed by the variation of the voltage going from the equilibrium of the open circuit to a weak or strong value according to the phase (charge or discharge), when an electric current passes through the battery.

These parameters depend on the meteorological characteristics of the site where the system is installed. In the following we will study the behaviour of the battery through these parameters.

Conclusion

In this chapter we have presented the photovoltaic system and its components which are the photovoltaic panels, the regulator, the inverter, the storage system and the loads. We then focused on the battery storage system in the photovoltaic system. We have highlighted its functioning and the different batteries available on the market. However, it is necessary to have an idea of its behaviour under certain conditions of use.

Chapter II

Moderation of accumulator batteries

Introduction

Moderation consists of translating the phenomena that occur in physical systems into a representation, often mathematical. It is therefore important to present the mathematical equations that allow us to model the phenomena taking place in the battery. Depending on the application and the constraints to which they are subjected, batteries react differently, so there is no single model that is accurate in all circumstances.

There are several electrical models of batteries in the literature. The choice of the model used for battery modelling is determined by the type of phenomena to be studied. In our work, we have studied three battery models allowing us to investigate several phenomena in storage batteries used in PV systems.

I- CIEMAT electrical model

The CIEMAT model can be considered simply to carry out an analysis of the various energy flows taking place inside and outside the battery system. 11 also allows the choice of the size of the system to be installed to be resolved [30].

1-1) CIEMAT electrical diagram

This model is based on the diagram in figure II.1, where the battery is described by two elements, a voltage source and an internal resistance, whose values depend on a number of parameters [10].

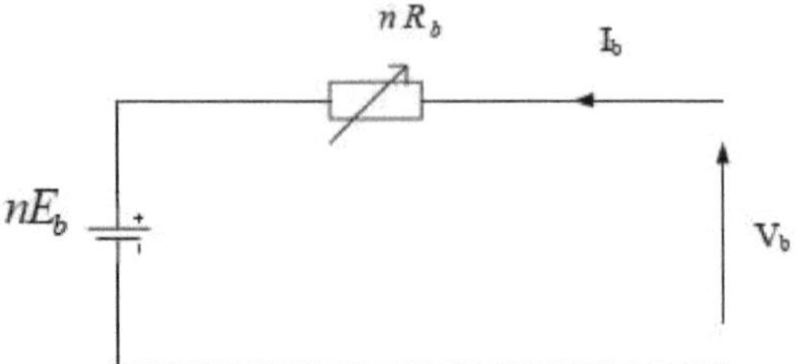

Figure II.1: Equivalent diagram of the CIEMAT model

The mathematical model describing the physical phenomena of charging and discharging based on Kirchhoff's laws is given by the following equation [30]: where *Vb* is the battery voltage, *Eb* is the electromotive force, *Rb* is the internal resistance, *Ib* is the battery current, *n* is the number of cells in the battery.

$$V_b = nE_b \pm nR_b I_b, \quad (1)$$

1-2) CIEMAT model equations

The CIEMAT model highlights several parameters of the battery taking into account some exogenous phenomena such as temperature:

❖ **Capacity**

In order to bring out the physical phenomena that govern the operation of the storage system through capacity, temperature should be taken into account. The capacitance model giving the amount of energy that can be released as a function of the average discharge current I_{moy} is given by the following equation [30]:

$$C_b = \frac{1.67C_t(1+0.005\Delta T)}{1+0.67(\frac{I_{moy}}{I_t})^{0.9}}, \quad (2)$$

with C_t the capacity of the discharged battery during a time t, /,1c the discharge current during the same time t, JTTthe heating of the battery.

The capacity model is established from the expression of the discharge current I_t for a period t, corresponding to the operating regime at C_t representing the capacity of the discharged battery for the period t, where ДT is the heating of the battery (assumed to be the same for all cells) in relation to the ambient temperature.

❖ **The state of charge**

The Cь capacity is used as a reference to determine the state of charge of the battery. We define Q as the amount of missing charge. We have the equation :

$$EDC = 1 - \frac{Q}{C_b} \tag{3}$$

The time evolution of the missing charge Q depends on the operating mode of the battery.

$$Q = I_b t \tag{4}$$

❖ **Discharge voltage**

The expression of the battery voltage is established from equations (1) and (2) which allow us to give a related structure of the internal elements of the battery as a function of the electromotive force, the internal resistance and the influence of parameters.

$$V_{bd} = n(1.965 + 0.12 \times EDC) - n\frac{|I_b|}{C_t} \times (\frac{4}{1+|I_b|^{1.3}} + \frac{0.27}{EDC^{1.5}} + 0.02) \times (1 - 0.007 \times \Delta T) \tag{5}$$

❖ **The charging voltage**

The voltage equation under load is similar to that obtained under discharge.

$$V_{bc} = n(2 + 0.16 \times EDC) + n\frac{|I_b|}{C_t} \times (\frac{6}{1+|I_b|^{0.86}} + \frac{0.48}{(1-EDC)^{1.2}} + 0.036) \times (1 - 0.025 \times \Delta T) \tag{6}$$

❖ **Battery resistance**

The internal resistance of the battery is not a constant value. It varies with the battery's state of charge, temperature and ageing. The internal resistance of the battery increases with its state of charge. Similarly, for a given state of charge, it increases as the battery ages. During discharge, the internal resistance is modelled by the equation :

$$R_d = \frac{1}{C_t} * \left(\frac{4}{1+|I_b|^{1.3}} + \frac{0.27}{EDC^{1.5}} + 0.02 \right) * (1 - 0.007 * \Delta T). \qquad (7)$$

During charging, the resistance of the battery is as follows:

$$R_c = \frac{1}{C_t} * \left(\frac{6}{1+|I_b|^{0.86}} + \frac{0.48}{(1-EDC)^{1.2}} + 0.036 \right) * (1 - 0.025 * \Delta T). \qquad (8)$$

EDC is the state of charge of the battery, Vbc and Vbd are the charging and discharging voltages of the battery, respectively, R_c and Rd are the internal resistances of the battery during charging and discharging, respectively, Ib is the current of the battery.

II- Thevenin's dynamic lineal model

Thevenin's dynamic linear model as shown in figure II.2 consists of an open circuit voltage Eo_c , an internal resistance Ro, and two parallel RC branches.

II-l) Schema equivalent

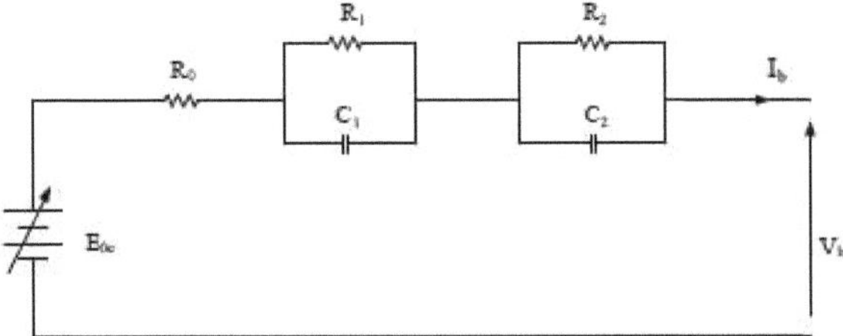

Figure II.2: Equivalent diagram of the Thevenin linear dynamic model

This model provides the transient response related to the double layer phenomenon of electrical polarisation and dynamic polarisation to the voltage Vb. The two arrays (R1//C1, R2//C2) react to two different time constants, n (fast) and T2 (slow) [13].

II-2) Model equations

❖ **State of charge of the battery**

The evolution of the state of charge of the battery is expressed by the following equation:

$$EDC = 1 - \frac{1}{C_n}\int I_b d\tau \quad (9)$$

with C_n the nominal capacity of the battery, 1ь the current of the battery.

❖ **Open circuit voltage**

The open circuit voltage E_{Oc} is a linear function of the state of charge and is given by the following equation:

$$E_{0c} = E_0 - K_e(1 - EDC) \quad (10)$$

Eo is the open circuit voltage when the battery is fully charged, K_e is a constant equal to 1.7

❖ **Resistance equations**

The resistances Ri and R2 of the model are expressed from the equations below:

$$DOC = 1 - \frac{1}{C(i_{avg})}\int I_b d\tau \quad (11)$$

$$R_1 = R_{10}\exp(-K_1(1 - EDC)) \quad (12)$$

$$R_2 = \frac{R_{20}}{DOC} \quad (13)$$

with C(iavg) the current dependent capacity of the battery. Elie is given by the constructed. DOC is the depth of charge of the battery, Ki is a constant equal to 7.

❖ Expression of battery voltage

The voltage Vb at the battery terminals determined by applying Kirchhoff's laws is expressed as follows:

$$\dot{V}_{C_1} = \frac{dV_{C_1}}{dt} = \frac{V_1}{R_1C_1} + I_b\frac{1}{C_1} \quad (14)$$

$$\dot{V}_{C_2} = \frac{dV_{C_2}}{dt} = \frac{V_2}{R_2C_2} + I_b\frac{1}{C_2} \quad (15)$$

$$V_b = E_{0c} - V_1 - V_2 - R_0I_b \quad (16)$$

with V_{C_1} et V_{C_2} the voltages at the terminals of capacitors Ci and C2 respectively.

Ill- Improved dynamic model of Thevenin III-l) Equivalent scheme

Shown in figure II.3, this model consists of a voltage source representing the electromotive force Eь, a resistance R_S d modelling the self-discharge, a resistor-capacitor network describing the polarisation effect and ideal diodes allowing to impose the direction of the current according to the regime (charge or discharge).

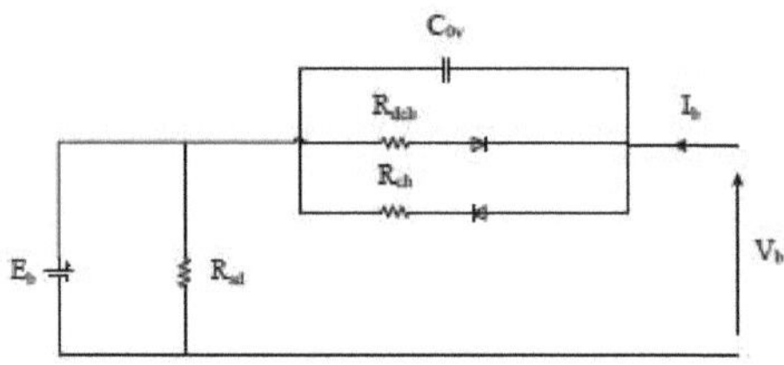

Figure II.3: Equivalent scheme of the improved dynamic model of Thevenin[4J

This model takes into account most of the non-linear phenomena of the battery cells during charging and discharging as well as their dependence on the state of charge of the battery; such as internal losses due to self-discharge and the polarisation effect within the battery.

III-2) Equations of the improved dynamic Thevenin model

❖ State of charge of the battery

During the charge/discharge cycle, the dynamic parameters of the battery depend on the state of charge, the charge/discharge rate and the temperature.

The expression of the state of charge in percent is given by the following equation:

$$EDC = EDC_i + (\frac{1}{C_n}\int_0^t I_b d\tau)100 \quad (17)$$

with EDCi the initial state of charge in percent, C_n the nominal capacity of the battery in Ah.

As the battery may suffer premature degradation, a revolutionary estimate of not only the state of charge but also the voltage of the battery in question should be made in order to design a battery control and management system.

❖ Open circuit voltage

The open circuit voltage E_b is determined by a linear approximation and varies directly with the state of charge of the battery. Its expression is as follows:

$$E_b = 0.01375 \times EDC + 11.5 \quad (18)$$

❖ The self-charging resistance

The self-discharge is electrically modelled by a resistance placed in parallel with an electrical source of a voltage representing the e.m.f. of the battery. Using the curve fitting technique we obtain a quadratic polynomial function representing this resistance. R_s *d is the* self-discharge resistance of the battery.

$$R_{sd} = -0.039 \times (EDC)^2 + 4.27 \times EDC - 19.23 \quad (19)$$

❖ Expression of discharge voltage

The discharge voltage is expressed by the following equation:

$$V_{bd} = E_b - I_b R_{dch}(1 - \exp(-\frac{t}{R_{dch} C_{ov}})) \qquad (20)$$

with *Cov* the polarisation capacitance. The discharge resistance is given by :

$$R_{dch} = R_{bdi} + R_{bd} \qquad (21)$$

Rbdi shows the change in voltage from the electromotive force during the transient phase and is dependent on the discharge current.

$$R_{bdi} = 1.01 \times \exp(-2.21 \times I_b) + 0.24 \times \exp(-0.06 \times I_b) \qquad (22)$$

The resistance Rbd represents the variation of the internal resistance with the state of charge during the discharge period.

$$R_{bd} = 2.926 \times \exp(-0.042 \times EDC) \qquad (23)$$

❖ **Polarisation capacity**

The phenomenon of polarisation in the battery results in the variation of the open circuit balance voltage to a lower or higher value depending on the phase (charge or discharge) when a current of *lb* flows through the battery.

The polarization capacity C_{ov} is given by :

$$C_{ov} = \frac{t_{ov}}{5R_{bdi}} \qquad (24)$$

with tov the setting time

❖ **Expression of the charging voltage**

The expression for charging voltage is similar to that for discharging.

$$V_{bc} = E_b + I_b R_{ch}(1 - \exp(-\frac{t}{R_{ch} C_{ov}})) \qquad (25)$$

The equivalent internal resistance Rch during charging is the sum of two resistances:

$$R_{ch} = R_{bei} + R_{bc} \quad (26)$$

Where *Rbd* reflects the change in voltage from the electromotive force during the transient phase and is dependent on the load current and Rb_c represents the variation of the internal resistance with the state of charge. Its expression is given as a quadratic polynomial determined from the curve fitting method.

$$R_{bc} = 9.32\times10^{-5}\times EDC^{2} + 0.01\times EDC + 0.028 \quad (27)$$

IV- Methodology

In order to study the battery as a means of energy storage in a PV system, we took Eseka as a site for which temperatures were collected. Eseka is a small town in the Centre region of Cameroon. Eseka has the following geographical coordinates: 3?38'59" (3°38'59) NORTH latitude, 3.65 degrees and 1C46'0" (10^5 46'0) EAST longitude, 10.7667 degrees, and is located at an altitude of 247m above sea level.

The following map shows the location of the town of Eseka:

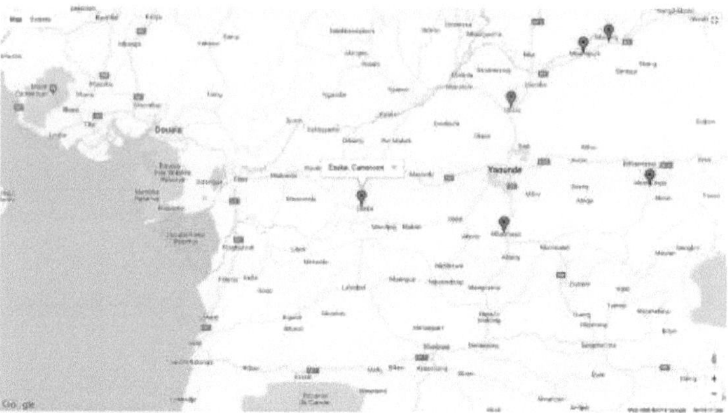

Figure II.4: Location map of Eseka

The meteorological data taken into account in our work are the maximum, minimum and average temperature of the Eseka site. These data presented in figure II.5 were collected over a period of 6 months from January to June with time intervals of 10 minutes.

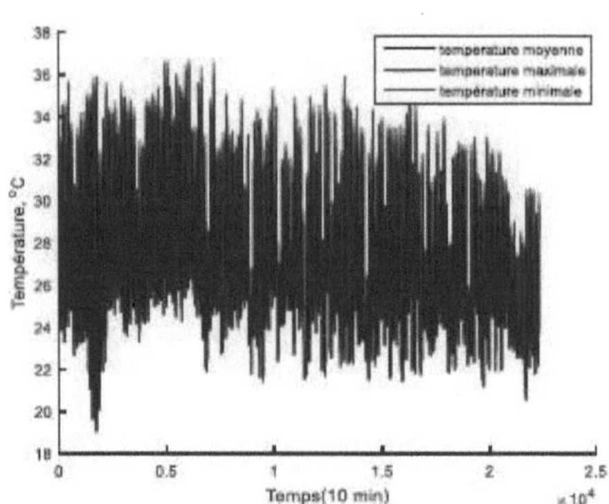

Figure II.5: Temperature variation at the Eseka site from January to June

In order to better present the variation of the different temperatures at the site, we show in figure II.6 the variations over each month. In the month of January a very small temperature variation is observed compared to the other months. The curves in figure II.6 (a) are plotted for the months of January, February, March and those in figure Π.6 (b) for the months of April, May, June. The data were collected from 18 January to 21 June. It can be seen that the temperatures at Eseka site oscillate very slightly between the maximum and minimum values.

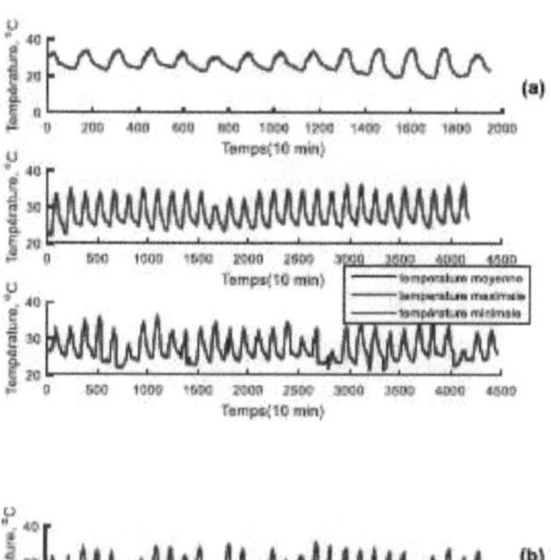

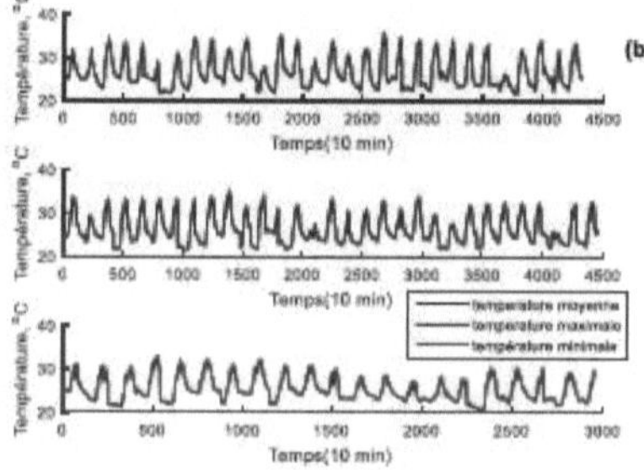

Figure II.6: Eseka temperature variation over 6 months

We recall that one of the objectives of this study is on the simulation techniques and the physical problems that take place in the battery. The simulation of the phenomena acting in the battery was done in the MATLAB environment.

MATLAB is a simple syntax matrix calculation software. In this work, we used Matlab (version 8.5) to simulate the behaviour of a lead-acid battery. Simulations are performed to study the battery parameters, their influence on the battery performance as well as the influence of exogenous phenomena such as temperature variation and load. The steps followed for the moderation and simulation of the battery models in this thesis are described in the following flow chart.

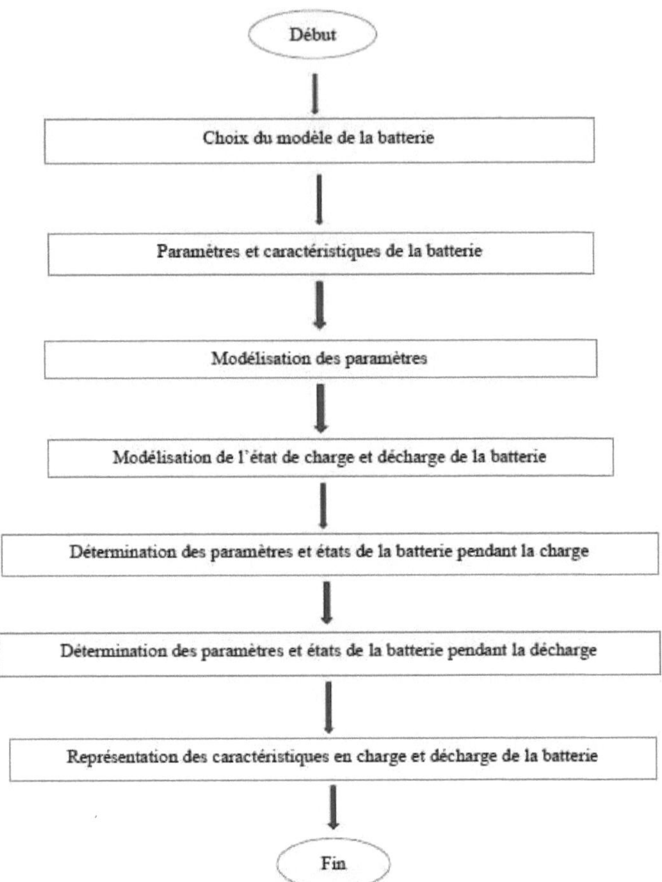

Figure II.7: Methodology used for moderation and simulation of lead acid batteries in this work

Conclusion

In this chapter we have presented some of the electrical battery models found in the literature. Each model presents different characteristics of the battery depending on the parameters to be studied. In addition, the methods used for moderation and further work are presented.

Chapter III

Simulation and results of some battery models

Introduction

In the previous chapter, three electrical battery models were presented. It is now a question of studying the behaviour of the battery via these models. For this purpose, a lead acid battery will be used. This is the most commonly used battery for energy storage in photovoltaic systems. Given the purchasing power in Cameroon, it is more appropriate because of its affordable market price and availability; it also has a good life span and reliability. In addition, the lead acid battery has a high nominative voltage of 2V per cell and a high energy efficiency. Thus, we will present the simulation results of these battery models studied in the Eseka site.

I- The CIEMAT model

For the simulation of this model, we have chosen a battery composed of 24 electrochemical accumulators of 2V in series. We have taken the capacity of the battery to be $C_t = 92$ Ah, and $C_t = 10I_t$.

1-1) Influence of Temperature

Temperature directly affects electrochemical phenomena by influencing ion mobility (solution conductivity and transfer kinetics). When the storage is in use, the temperature T (°C), which is very high or low, can strongly influence the functioning of the battery. Thus, we studied the influence of the maximum (36.73°C), minimum (19°C) and average (26.89°C) temperature of Eseka on the capacity, state of charge, charge voltage and discharge voltage of the battery.

> Influence of temperature on battery capacity

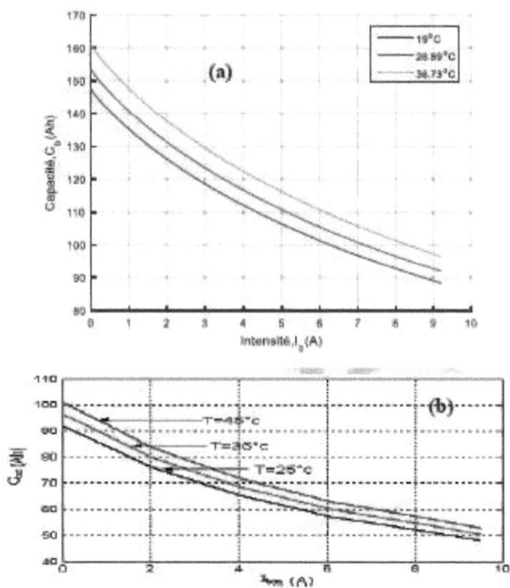

Figure III.l: Variation of capacitance with intensity for different temperatures different temperatures

Figure 111.1(a) shows the simulation results obtained in our work while Figure 111.1(b) shows the results obtained by Idir ISSAD et al. in their work.

Figure 111.1(a) shows the change in battery capacity as a function of current intensity for different site temperatures. We can see that the capacity increases with temperature. For temperatures lower than 27°C there is a significant reduction in battery capacity while higher temperatures produce a slight increase in capacity, but according to [24] this increases water loss and decreases the life of the battery.

life of the battery. We note that the results obtained for our work at the Eseka site are in line with the literature.

> Influence of temperature on the state of charge

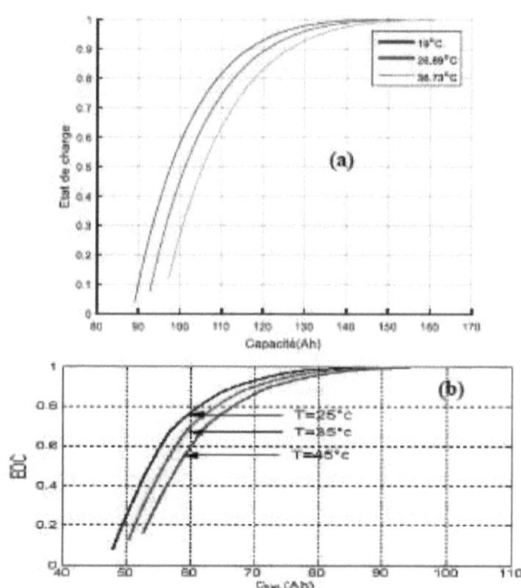

Figure III.2: Variation of the state of charge as a function of the capacity

Figure 111.2(a) shows the simulation results obtained in our work, while Figure 111.2(b) shows the results obtained in the work of Idir ISSAD et al.

Figure 111.2(a) shows the revolution of the battery's state of charge as a function of its capacity for different temperatures. It can be seen that the higher the temperature, the lower the state of charge of the battery. Similarly, when the battery is subjected to a low

In the same way, when the battery is subjected to a low temperature, it is charged more quickly than when it is subjected to a high temperature.

> Influence of temperature on the state of discharge

Figure III.3 shows the evolution of the battery state of charge as a function of capacity for different temperatures at the test site.

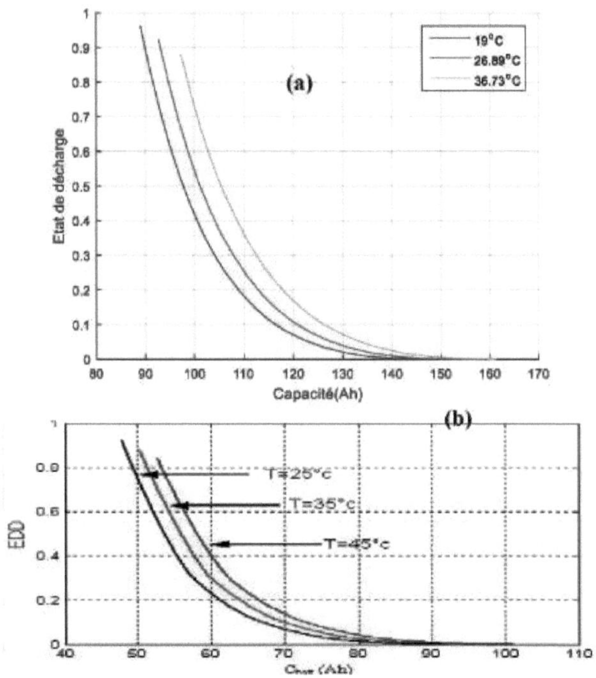

Figure 1 I.3 : Variation of the state of charge as a function of the capacity

Figure 111.3(a) shows the simulation results obtained in our work, while Figure 111.3(b) shows the results obtained by Idir ISSAD et al. In the case of decharge, the decreasing temperature is accompanied by a decreasing state of decharge.

> Influence of temperature on charging voltage

Figure III.4 plots the variation of the battery charging voltage against the current for different temperatures. It can be seen that as the site temperature increases, the voltage during charging decreases. This variation is even more important with the value of the current. Figure 111.4(a) shows the variation of the battery voltage during charging for a variable capacity. This variation can be due to the temperature variation and also to the ageing phenomenon. Figure 111.4(b) is plotted for a fixed capacity of 160Ah.

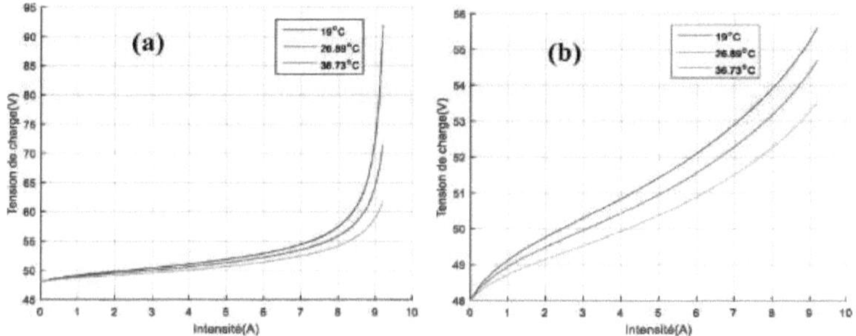

Figure III.4: Variation of the load voltage with current intensity

> Influence of temperature on the discharge voltage

Figure III.5 shows the variation of the battery discharge voltage as a function of current intensity for different temperatures.

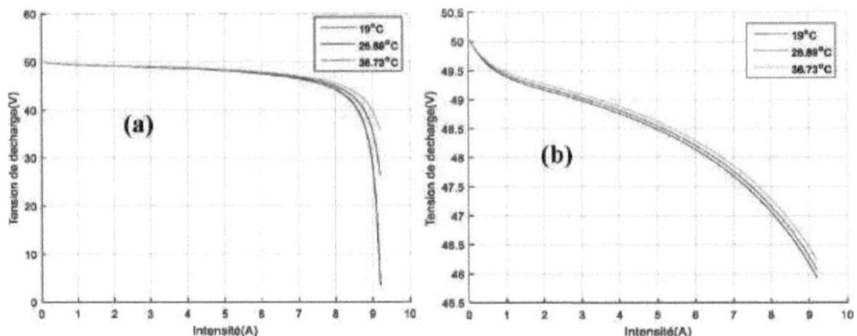

Figure III.5: Variation of the discharge voltage as a function of the current intensity

It can be seen that the temperature has a remarkable influence on the discharge voltage. The higher the temperature, the higher the discharge voltage. This variation is even more visible when the intensity of the discharge current is increased. Figure 111.5(a) shows the variation of the battery voltage during discharge for a variable capacity. This variation can be due to the temperature variation and also to the aging phenomenon. Figure III.5(Ь) is plotted for a fixed capacity of 160Ah.

Using equation (28) to determine the rate of influence of temperature on the voltages during charging and discharging of the battery, we find that the temperature at the Eseka site has a small influence on the battery, i.e. 1.9791% during charging and 0.2885% during discharging.

$$M = \left(\frac{1}{N}\sum_{i=0}^{N}\frac{(V_{T_{max}} - V_{T_{min}})}{V_{T_{ref}}}\right)*100 \qquad (28)$$

With $V_{T_{max}}$, $V_{T_{min}}$, $V_{T_{ref}}$ respectively the voltages for the maximum, minimum and reference temperatures and N the number of samples.

1-2) Influence of time

Time is a crucial variable for the study of the behaviour of any physical system. In this subsection, we present the influence of time on the parameters of the storage battery at the Eseka site.

> Influence of time on the state of charge

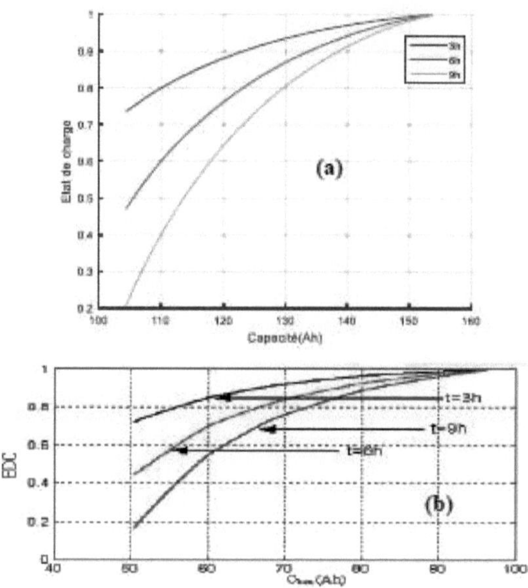

Figure III.6: Evolution of the battery state of charge as a function of capacity for capacity for different values of time

Figure 111.6(a) shows the simulation results obtained in our work while Figure 111.6(b) shows the results obtained by Idir ISSAD et al. in their work.

Figure III.6 shows the evolution of the battery's state of charge as a function of capacity for different values of charging time. We can see from Figure 111.6(a) that the state of charge increases with battery capacity. This figure also shows that the variation of the state of charge as a function of the capacity

This figure also shows that the variation of the state of charge as a function of capacity is strongly dependent on time; as time increases, the value of the state of charge decreases. This is in line with the results of the literature.

> Influence of time on the state of discharge

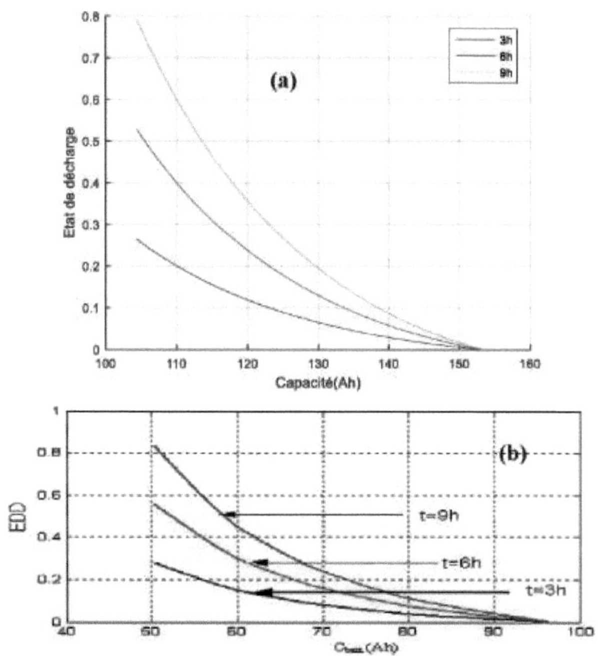

Figure III.7: Evolution of the state of discharge of the battery as a function of capacity for different values of time

Figure 111.7(a) shows the simulation results obtained in our work, while Figure 111.7(b) shows the results obtained in the work by Idir ISSAD et al.

Figure 111.7(a) shows that the state of discharge decreases as the capacity of the battery increases. It can be seen that for a long discharge time the value of the state of discharge increases considerably, which is in line with the results in the literature. This suggests that it would be wise to choose a battery with a good capacity that would maintain a state of charge greater than the depth of discharge after a certain number of hours of use in discharge.

> Influence of time on charging voltage

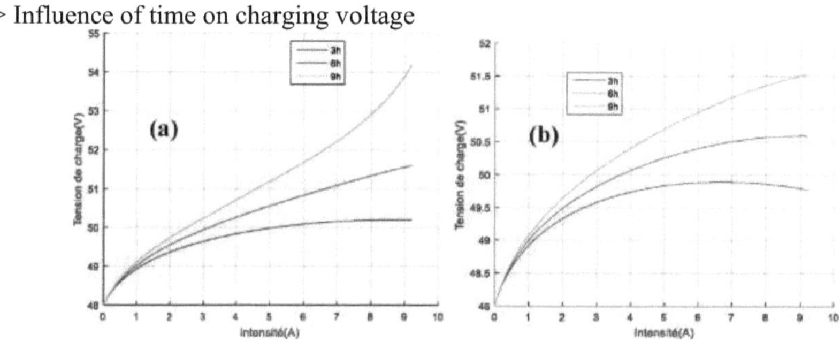

Figure III.8: Evolution of the discharge voltage as a function of the current

According to figure III.8, the charging voltage is very variable with time. For a high charging time the voltage increases. This variation is all the more important as the intensity of the charging current increases. This means that the higher the current, the faster the battery is charged. Figure 111.8(a) shows the variation of the battery voltage during charging for a variable capacity. This variation can be due to the temperature variation and also to the ageing phenomenon. Figure 111.8(b) is plotted for a fixed capacity of 160Ah. It can be seen that the charging voltage of a battery with increasing capacity is greater than that of a battery with a fixed capacity

> Influence of time on the discharge voltage

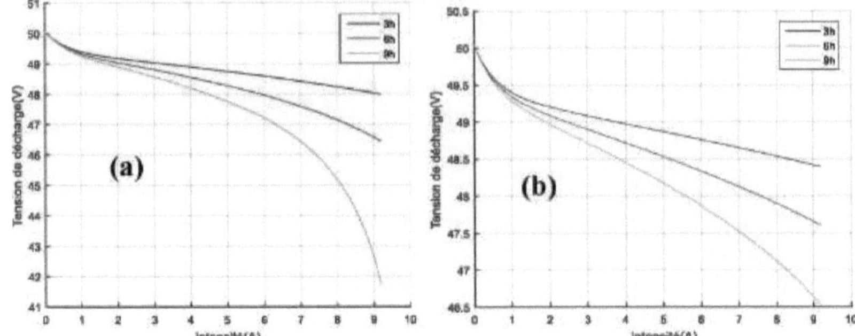

Figure III.9 : Evolution of the discharge voltage as a function of time

Figure III.9 shows that the discharge voltage decreases with time, i.e. the longer the discharge time, the lower the voltage. This variation is well observed for high current intensities. Figure 111.9(a) shows the variation of the battery voltage during discharge for a variable capacity. This variation may be due to the temperature variation and also to the ageing phenomenon. Figure 111.9(b) is plotted for a fixed capacity of 160Ah. It shows that during discharge the voltage of a battery with a varying capacity is lower than that of a battery with a fixed capacity.

1-3) State of charge of the battery over time

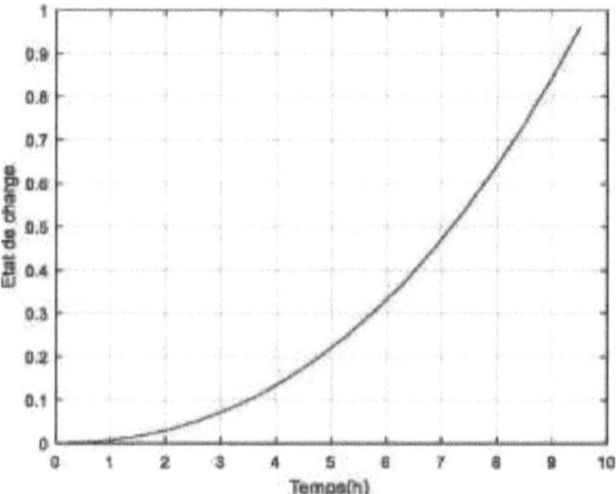

Figure 10 . 10: State of charge as a function of time

Figure III.10 shows the variation of the state of charge over time. It can be seen that the battery is gradually charged over time.

1-4) Internal resistance

The internal resistance of a battery is the sum of several elemental resistances representing various phenomena observed in electrochemical studies and which are used for its design and manufacture.

> **Internal load resistance**

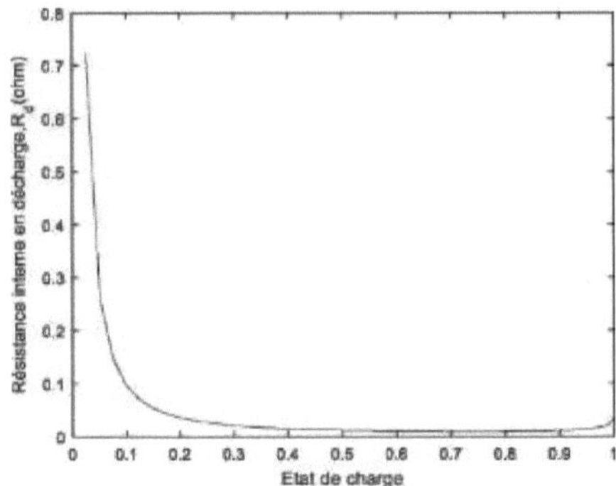

Figure 11 . ll: Internal resistance of the battery during discharge

Figure III.ll shows the change in internal resistance during discharge as a function of the state of charge of the battery. It can be seen that when the battery is charged, the internal resistance is low and it becomes high at a state of charge close to zero.

> **Internal resistance under load**

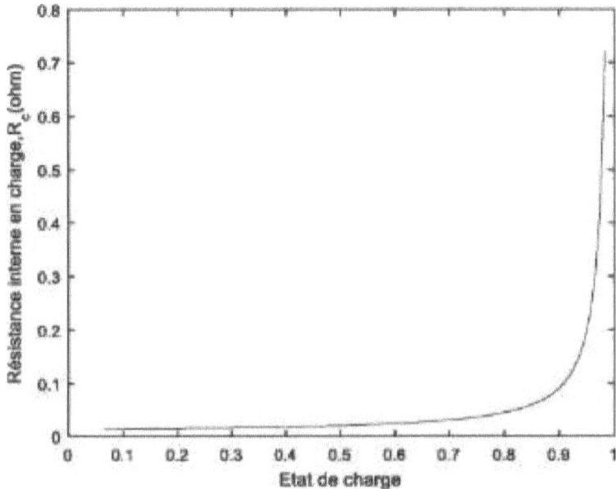

Figure 12 . 12: Internal resistance of the battery during charging

Figure III.12 shows the variation of the internal resistance of the battery during charging as a function of the state of charge. It can be seen that the value of this resistance becomes more important when the full charge is reached. This phenomenon (variation of the internal resistance of the battery during charging and discharging) is due to the increase of the resistance of the electrolyte and mainly of the electrodes.

1-5) Profile of the battery state of charge as a function of temperature and time

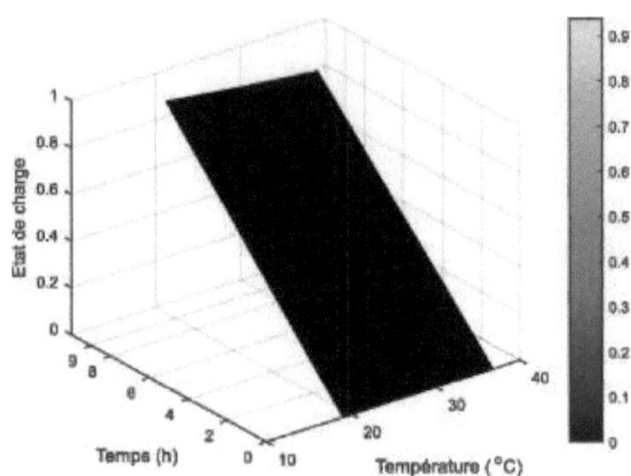

Figure 13 . 13: Profile of the battery's state of charge as a function of temperature and time

Figure III.13 shows the evolution of the state of charge of the battery over time for various temperatures. The coloured line represents the variation of the state of charge. It can be seen that during the charging of the battery, its state of charge increases over time. However, during this process, for a simultaneous increasing temperature change, the battery tends to decrease its state of charge; the battery is thus charged more slowly.

1-6) Internal resistance profile of the battery during charge and discharge

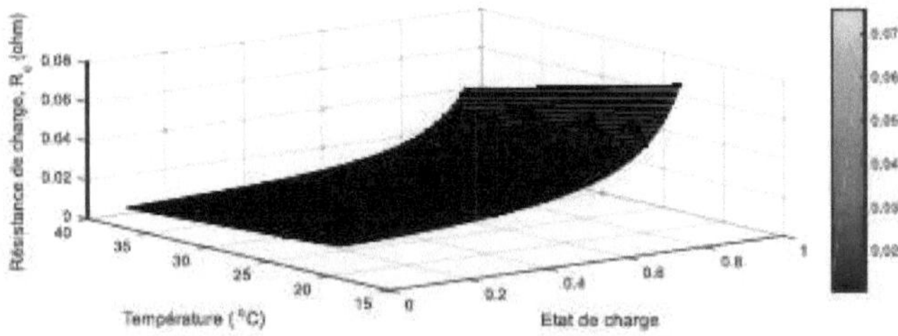

Figure 14 .14: charging resistance profile as a function of temperature and battery state of charge

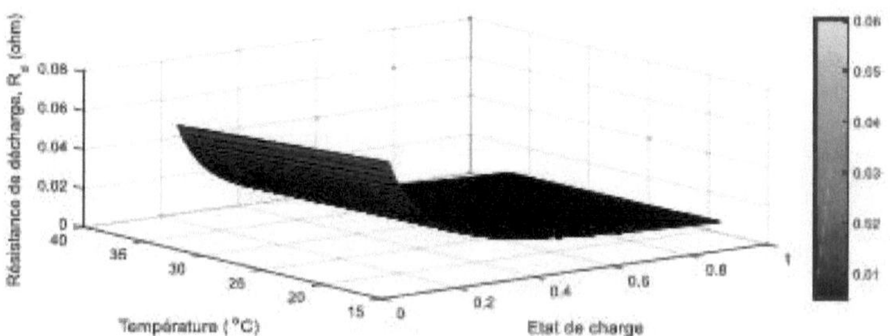

Figure 15 .15: Discharge resistance profile as a function of temperature and battery state of charge

It can be seen from figures III.14 and III.15 that the resistance of the battery changes inversely with temperature. During charging, an increase in temperature leads to a decrease in resistance. In fact, the internal resistance of the battery varies (increases) with the state of charge, an increase in temperature simultaneously contributes to lowering it and thus to slowing down its evolution.

During discharging, as the state of charge of the battery decreases, the internal resistance increases. A simultaneous increase in temperature induces a decrease in resistance which tends to slow down its evolution. The coloured line represents the variation of the internal resistance of the battery.

II- Thevenin's dynamic lineal model

We have performed the simulation of the Thevenin linear dynamic model that we present in this subsection for: Ki = 7; Rio = 0.160 during loading and Rio = 0.020 during unloading; R20⁼ 0.00550; Ro = 0.120 during loading andRo = 0.0570 during unloading.

II-l) Discharge voltage

Figure III.16 shows the evolution of the discharge voltage over time.
It can be seen that this voltage decreases with increasing discharge time.

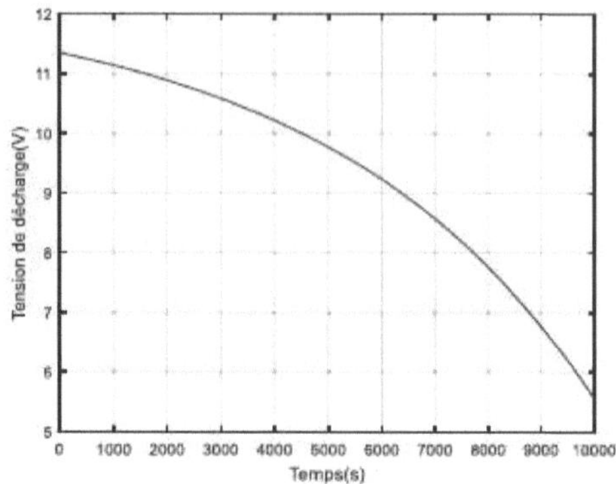

Figure III. **16: Evolution of the discharge voltage as a function of time**

II-2) Charge voltage

Figure III.17 shows the involution of the charge voltage over time.

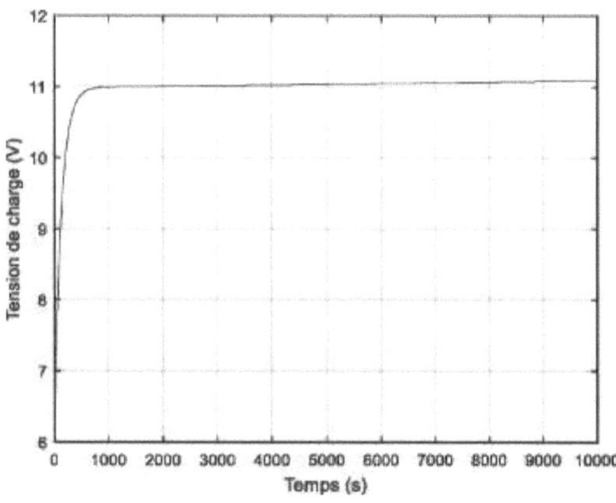

Figure 16 . 17: Evolution of the charge voltage as a function of time

It can be seen that the rate of charge (voltage increase) of the battery is rapid at the beginning of the charge and then becomes almost constant. When the state of charge exceeds 80-90%, voltage regulation takes place to limit the rate of charge. The excess energy that would be supplied to the battery is dissipated in the form of heat and in the electrolysis of the water contained in the electrolyte.

II-3) Variation of resistances

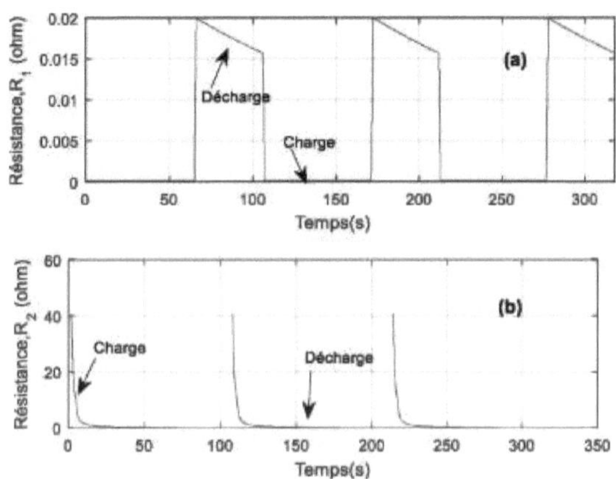

Figure 17 . 18: Evolution of resistances Rl(a) and R2(b) during the charge/discharge cycle of the battery

Figure III.18 shows the variation of resistances during the charge and discharge cycle of the battery. A small variation (increase) of the resistance Ri can be seen during charging, while during discharging, the variation (decrease) is significant. On the other hand, the resistance R2 follows the opposite process.

II-4) Voltage and current variations in charging and discharging

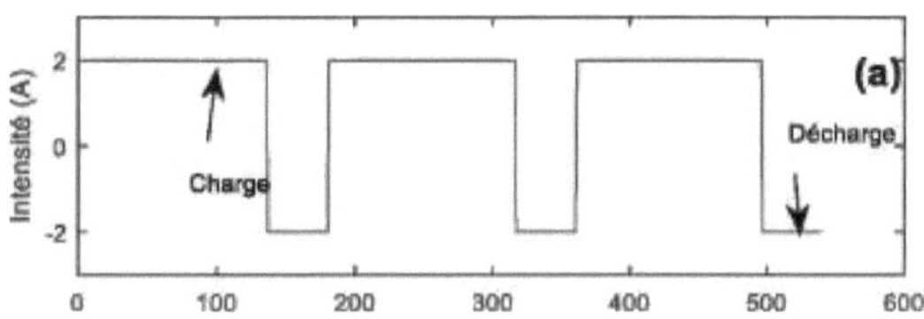

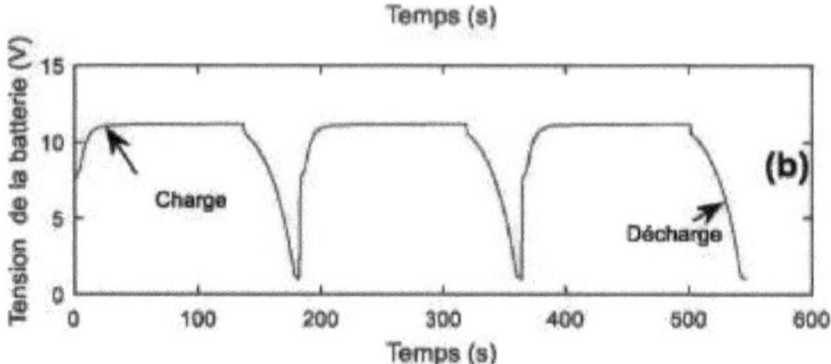

Figure 18 . 19: Evolution of battery current(a) and voltage(b) as a function of time

Figure III.19 shows the variation of the current (a) and voltage (b) during the charge/discharge cycle of the battery. The discharge current is counted negatively, while the charge current is counted positively. It can be seen that for the same current value, the charging time is longer than the discharging time. During charging, three zones are observed: the first zone is represented by a sudden (linear) increase in voltage at the beginning of the charge, the second is represented by a non-linear increase in voltage and finally the half-zone is represented by an almost constant variation in the battery voltage. During discharging, there is a sharp (linear) decrease in voltage at the beginning followed by a non-linear decrease.

II-5) Influence of capacity

> Influence of capacity on battery discharge voltage

Figure III.20 shows the variation of the voltage during discharge as a function of time for different values of the capacitance.

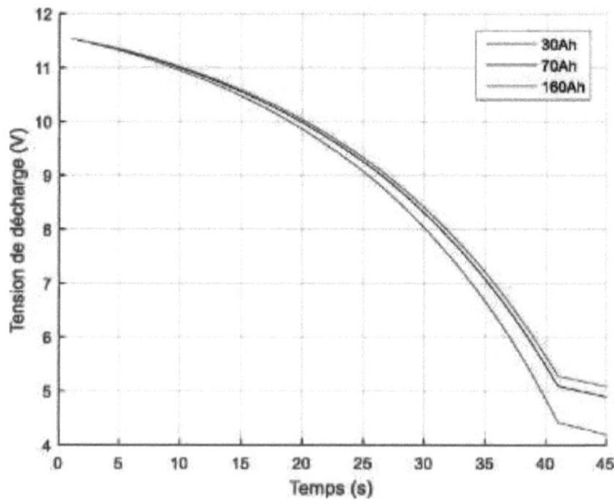

Figure III.20: Evolution of the battery discharge voltage as a function of time for different capacities

Figure III.20 above shows that during discharging, the battery voltage decreases as the battery capacity decreases over time.

> Influence of capacity on battery charging voltage

Figure III.21 shows the variation of the battery voltage during charging over time for different values of capacity

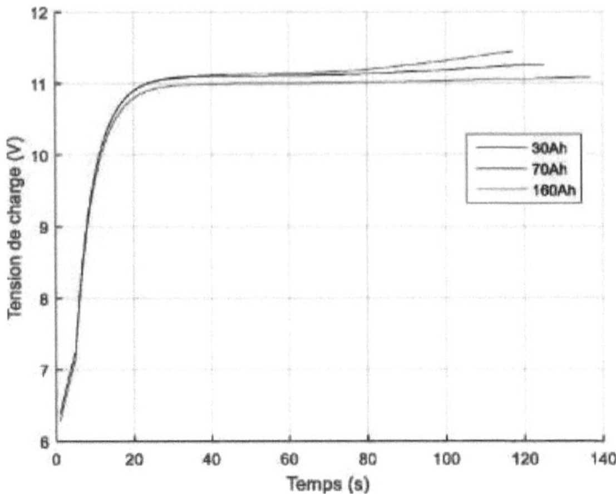

Figure III.21: Evolution of the battery charge voltage as a function of time for different capacities

It can be seen that as the battery capacity increases, the voltage of the battery during discharge increases less and less rapidly in the linear area.

III- Thevenin's improved dynamic model

We have performed the simulation of the improved Thevenin dynamic model that we present in this section for: t_{ov}= Imin; Rbci= 1.50.

III-l) Variation of the electromotive force

Figure III.22 shows the evolution of the electromotive force as a function of the state of charge of the battery.

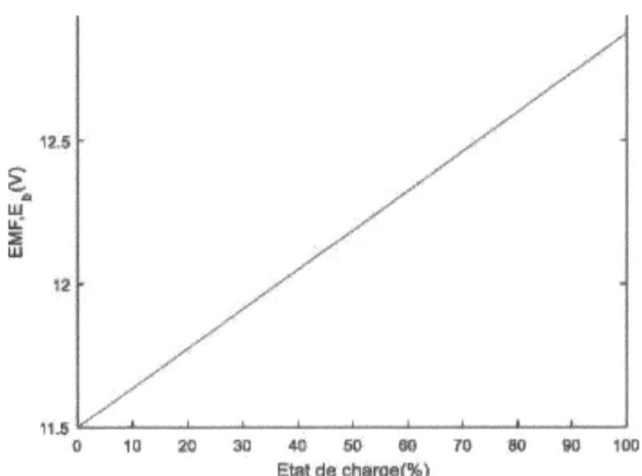

Figure 22 .. 22: Variation of the electromotive force as a function of the state of charge

The electromotive force is a linear function of the state of charge of the battery as shown in Figure III.22. The more the battery is charged, the greater its electromotive force.

III-2) Self-charging resistance

Figure III.23 shows the variation of self-charging resistance as a function of the state of charge of the battery.

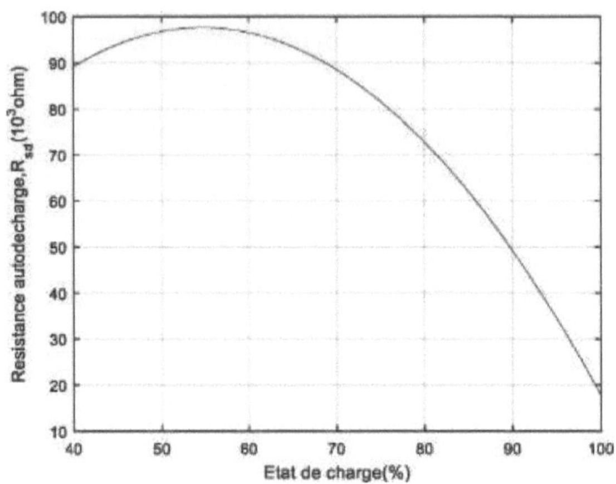

Figure 23 . 23: Variation of self-charging resistance as a function of state of charge

It can be seen that for a state of charge between approximately 40% and 53% the self-discharge resistance increases and begins to fall immediately thereafter. Figure III.23 shows that for an unused battery the self-discharge is more pronounced for a low state of charge and for a fully charged battery the self-discharge is less important. Thus, a lightly charged battery has a high self-discharge capacity.

III-3) State of charge

Figure III.24 (a) shows the revolution of the state of charge of the battery at rest as a function of time.

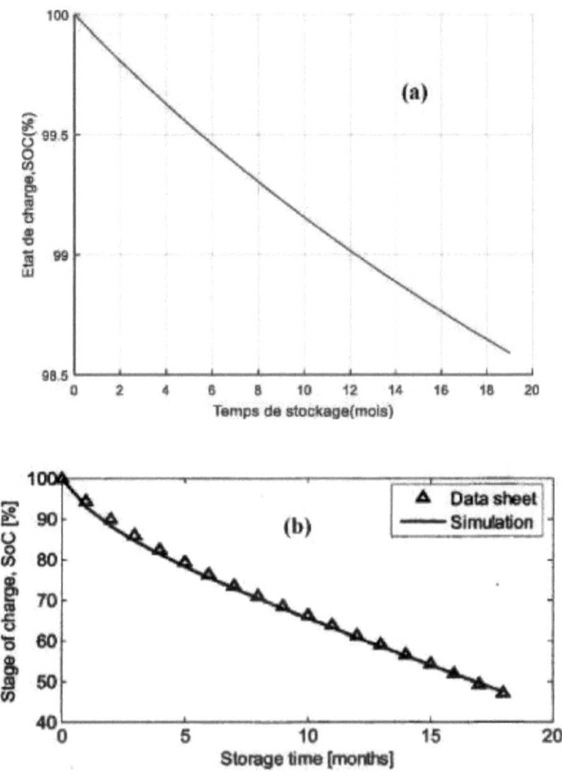

Figure 24 . . 24: State of charge of the battery as a function of time

Figure III.24 (b) shows the capacity loss due to self-discharge of the battery obtained by Jantharamin et al. in their work. A slow and gradual discharging over several months can be seen due to self-discharge of the battery. Thus we find that the capacity of a battery tends to decrease over time even when it is not discharging. The results found are in line with the literature.

III-4) Resistance during discharge

> Evolution of the Rbdi resistance

Figure III.25 shows the variation of the resistance Rbdi as a function of the discharge current

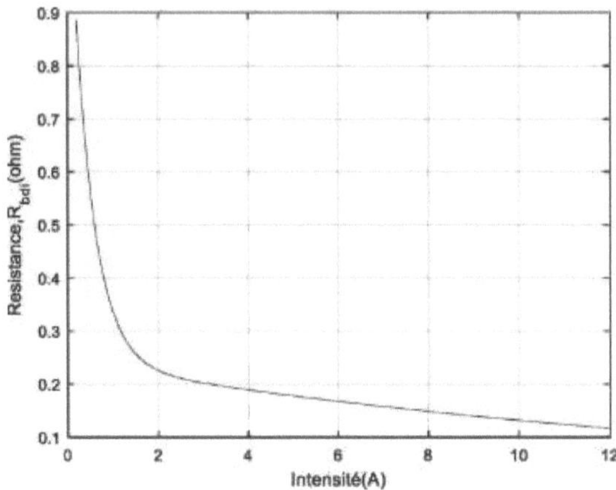

Figure III.25 : Variation of the resistance Rbdi as a function of the discharge current

The resistance Rbdi decreases during the discharge, which illustrates the variation of the voltage with the electromotive force.

> Evolution of the resistance Rbd

Figure III.26 shows the variation of the resistance Rbd as a function of the state of charge of the battery.

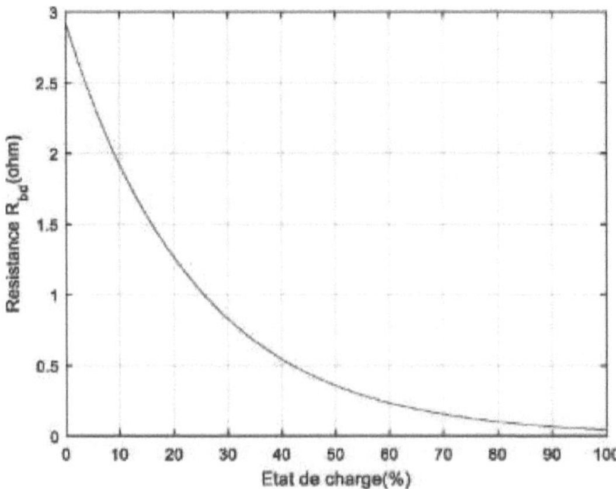

Figure 26 . 26: Variation of the resistance Rbd as a function of the state of charge

It can be seen that the internal resistance of the battery increases as the battery is discharged.

III-5) Discharge voltage

Figure III.27 shows the involution of the battery voltage during discharge as a function of time.

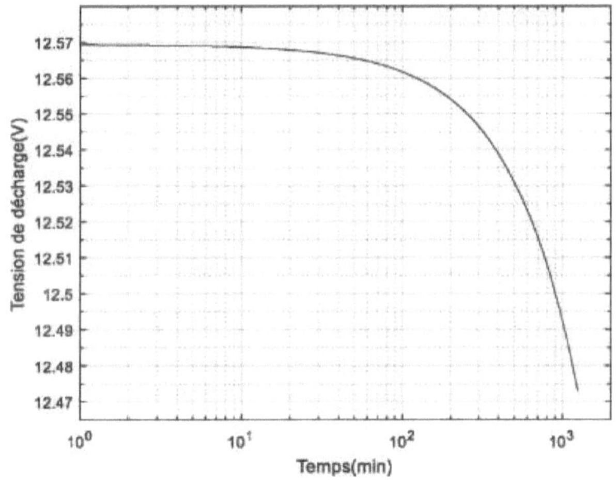

Figure 27 . 27: Evolution of the discharge voltage as a function of time

We plot the evolution of the battery voltage during discharge over time for a current of 0.5A and a capacity of 160Ah. It can be seen that the discharge voltage decreases with the discharge time which is in agreement with the results obtained for the previous models.

III-6) Evolution of the Rbc resistance

Figure III.28 shows the variation of the resistance Rbc as a function of the state of charge of the battery.

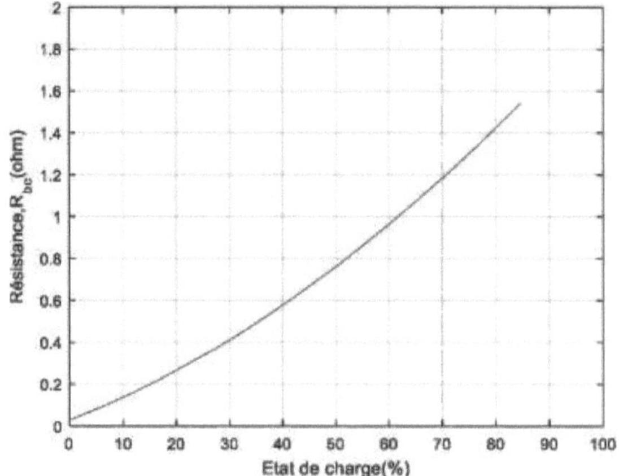

Figure 28 . 28: Variation of the resistance Rbc as a function of the state of charge

The resistance Rbc represents the internal resistance of the battery during charging. It can be seen that the internal resistance during charging of the battery increases with the state of charge.

III-7) The charging voltage of the battery

Figure III.29 shows the involution of the battery voltage during charging as a function of time.

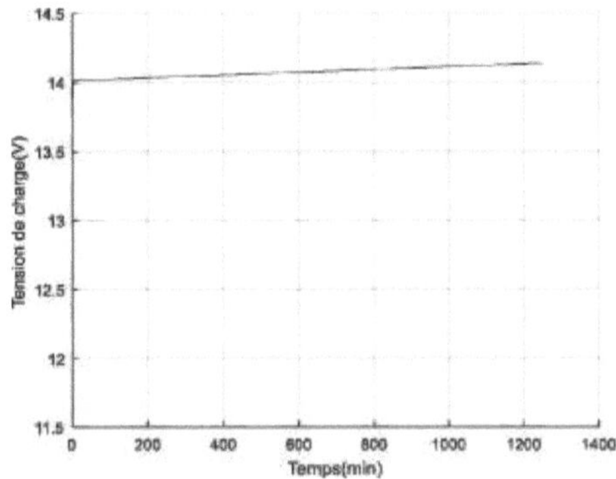

Figure 29 . . 29: Battery charge voltage versus time

There is a sudden increase in voltage at the beginning of the charge, followed by a progressive increase.

III-8) Influence of capacity on charge and discharge voltages

Voltages are plotted against time for different battery capacities with a current of 2A during the charging and discharging phases.

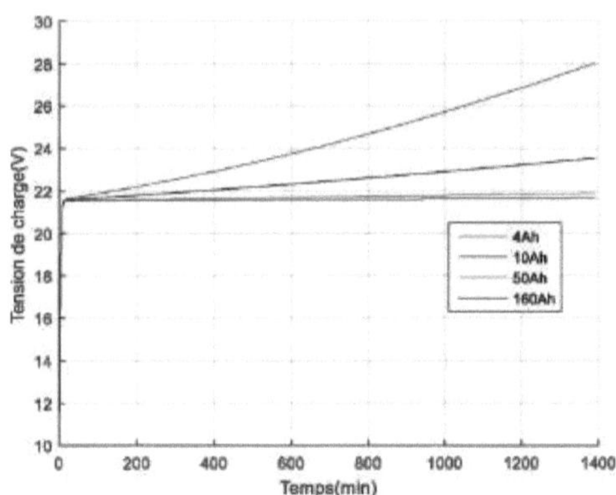

Figure 30 . 30: Influence of capacitance on voltage during charging

Figure III.30 shows a sharp increase in voltage over time at the start of charging, with voltages rising with capacity. Charging is faster for a battery with a low capacity, whereas for large capacities there is a small increase in voltage. A decrease in the capacity of the battery leads to an increase in resistance and thus an increase in the voltage of the battery.

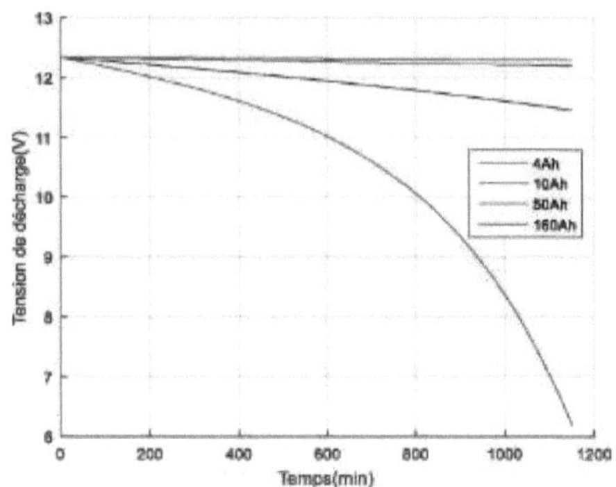

Figure 31 . 31: Influence of capacitance on voltage during discharge

It can be seen from figure III.31 that the lower the capacity, the lower the battery voltage during discharge. A battery with a low capacity discharges faster than one with a high capacity.

III-9) Influence of the current intensity on the charge and discharge voltages

The influence of the current intensity on the charging and discharging voltages is studied for a battery with a capacity of 160Ah.

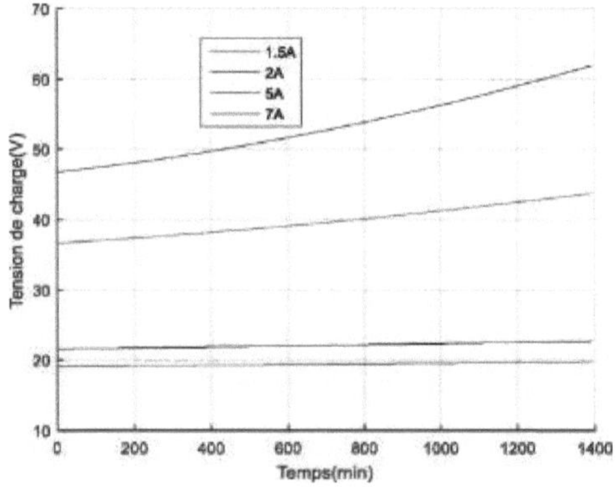

Figure 32 . 32: Influence of the current on the voltage during charging

Figure III.32 shows that for a high current the voltage increases rapidly over time and is much higher than that of a battery with a low current.

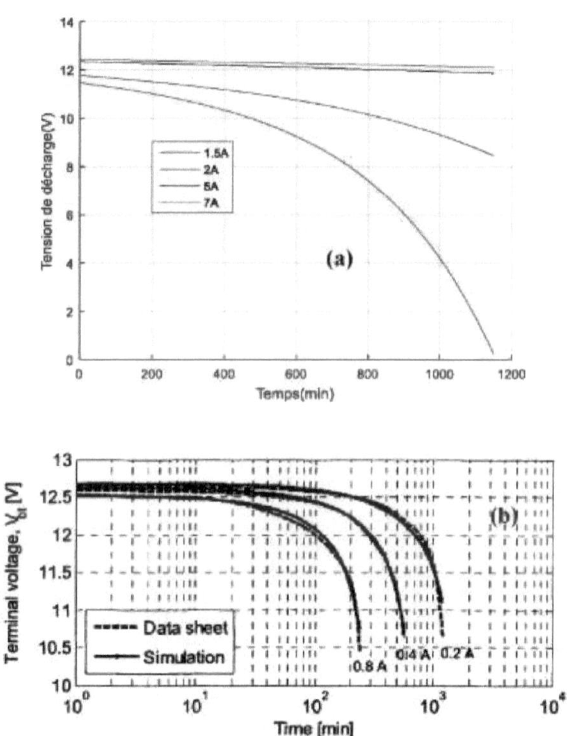

Figure 33 . 33: Influence of the current on the voltage during discharge

It can be seen from Figure 111.33(a) that during discharge the variation is more pronounced at high intensities and varies very little at low intensities. Thus, a battery discharges faster at high currents. Figure 111.33(b) shows the result obtained by Jantharamin et al. in their work. It can be seen that the voltage profiles obtained in our work are in line with the literature.

IV- Summary

As illustrated in the previous sections, we have studied the behaviour of the lead acid battery by modelling it using three different electrical models. Indeed, each model used has different characteristics.

This section is devoted to a summary of the three models studied in order to highlight the advantages of each. The models discussed above deal with the state of charge, voltage and internal resistance of the battery.

We found that the internal resistance increases as the battery is discharged. The internal resistance of the battery varies slightly for the CIEMAT model (for a state of charge of 80% we have 0.050 in charge and 0.010 in discharge) while we observe a strong variation in the improved dynamic model: for a state of charge of 80% we have 1.40 in charge and 0.20 in discharge. This difference is a function of the model studied, hence the interest in studying several models. An experimental validation is therefore necessary to determine the appropriate model to study the variations of the internal resistance of the lead acid battery. We found that the voltage during charging of a battery modelled with the CIEMAT model increases and becomes almost constant over time, while for the improved dynamic Thevenin model it shows a sharp increase up to a value of 12.5V before gradually increasing. Similarly, during discharging, the battery voltage decreases progressively for the CIEMAT model and the Thevenin improved dynamic model. In contrast, for the dynamic linear model, the voltage profile during the charge/discharge cycle shows a sharp increase and decrease at the end of the charge and discharge respectively. The specificity of each of the three models studied leads us to understand that there is no single model to study the functioning of a battery.

Conclusion

In this chapter, we have studied the behaviour of the battery by means of three models, namely the CIEMAT model, the Thevenin dynamic linear model and the Thevenin improved dynamic model. The evolution of different battery characteristics such as state of charge, charge and discharge voltages and internal resistances were highlighted. Each of the models allowed the study of different phenomena taking place in the battery.

General conclusion

The demand for electricity is growing rapidly and the means used to meet it are still insufficient. Photovoltaic energy is a solution to this problem. However, the energy produced by the sun is intermittent and requires a storage system for a photovoltaic installation. In this context, this work is devoted to the modelling and simulation of some battery models used in photovoltaic systems. In fact, energy storage by means of batteries has the advantage of being less expensive and accessible to Cameroonians given their purchasing power.

Therefore, in Chapter 1, the photovoltaic system was first presented through its components and the different types of installation possible, while focusing on the storage system. In this chapter we also presented the different types of battery modelling used in photovoltaic systems, namely empirical modelling, chemical modelling and electrical modelling. Then, in chapter 2 we studied three electrical models of the battery, namely the CIEMAT model, the Thevenin dynamic linear model and the Thevenin improved dynamic model. Each of these models presented particularities in determining the behaviour of the battery under certain conditions of use. Finally, Chapter 3 presented the results of the simulations of these three models of the battery assumed to be implemented in the Eseka site, Central Region of Cameroon. These results focus on specific parameters such as state of charge, charge and discharge voltages. We also studied the influence of the temperature of the operating site and the weather on these parameters. The CIEMAT model allowed us to highlight the influence of exogenous parameters such as the temperature of the site used, following which we found that the Eseka temperature has very little influence on the battery because this temperature itself varies very little; while the Thevenin dynamic linear model allowed us to express the change in battery voltage over time and finally the Thevenin improved dynamic model allowed us to highlight the self-charging phenomenon. As a result of these analyses, we note that the charging and discharging current intensities have an influence on the battery. Too high currents can contribute to the deterioration of the battery, as can too high or low temperatures. We have also found that the self-discharge phenomenon increases as the battery ages.

As solar batteries are not manufactured in Cameroon and even less under its meteorological conditions, it is imperative to determine the optimal conditions of use of these batteries. This is done by modelling these batteries under local conditions as done in this work. This is how the perspectives of our work take shape, it will be for us to determine the influence of self-charging on the life of the battery and to determine the conditions of use of the h-icon battery to optimize its capacity and its life span.

References

[1] Xiangjun Li ;Hui Dong; Lai Xiaokang, "Battery Energy Storage Station (BESS) -Based Smoothing Control ofPhotovoltaic (PV) and Wind Power Generation Fluctuations." pp. 464-473, 2013.

[2] N Jantharamin;L Zhang, "A New Dynamic Model for Lead-Acid Batteries." pp. 86-90, 2014.

[3] James D. Maclay; Jacob Brouwer; G. Scott Samuelsen, "Dynamic modeling ofhybrid energy storage systems coupled to photovoltaic generation in residential applications." lip. 2007.

[4] Y. K. Adel El shahat, Rami Haddad, "Lead Acid Battery Modeling for PV Applications." 2015.

[5] H. L. Chan; D. Sutanto, "A new battery model for use with battery energy storage systems and electric vehicles power systems." 2000.

[6] M. E. Glavin, P. K. W. Chan, S. Armstrong, W. G. Hurley, and I. Fellow, "A Stand-alone Photovoltaic Supercapacitor Battery Hybrid Energy Storage System." 2008.

[7] Boumedienne Fellah, "Hybrid photovoltaic-wind system for electricity production. Application aux sites de Tlemcen et de Bouzareah." Master's thesis in physics, Universite Abou-Bakr Belkaid de Tlemcen, 2012.

[8] M. B. BOUGHARI Imad, "Etude et optomation d ' un systeme photovoltaique." Memoire de master de physique, Universite Kasdi Merbah Ouargla, 2016.

[9] Africa Renewable Energy Access Programme, Solar Photovoltaic Energy System for Community Facilities and Services. 2010.

[10] Mazouz Karim, "Study of a storage battery in a photovoltaic system." Master's thesis in physics, Universite Abderrhamane-Mira-Bejaia, 2012.

[11] Mambrini Thomas, "Caracterisation de panneaux solaires photovoltaiques en conditions reelles d ' implantation et en fonction des differentes technologies." 194p. 2015.

[12] R. A.-P. T9-13:2013, PV-diesel hybrid mini-grids for rural electrification. 2013.

[13] Aissa BOUTTE, "Identification of internal parameters of a battery for photovoltaic applications." These de doctorat en sciences, Universite des sciences et de la technologic d'Oran Mohamed, 2015.

[14] P. T. Jeremie Ciceron, Arnaud Badel, "Superconductor energy storage and the S3EL electromagnetic launcher." 2016.

[15] Electric Company Nippon and United States, "Supercapacitor." pp. 33-70.

[16] Fabrice Delfosse, "Determination of the state of charge of batteries in an electric vehicle. University of Liegem 1998.

[17] Singo Akassewa Tchapo, "Photovoltaic power system with hybrid storage for energy autonomous housing." These de doctorat, Universite Henri Poincare, Nancy-I, 2010.

[18] B. A. H. BenMulton, "Stationary electric energy storage: why and how?" 2012.

[19] Simpore Sidiki, "Modelisation , simulation and optimisation of a compressed air storage system coupled to a building and a photovoltaic production." These de doctorat, Universite de laReunion, 2019.

[20] E. Korsaga et al, "Comparison and determination of appropriate storage devices for an autonomous photovoltaic system in the Sahelian zone." 25p, 2018.

[21] B. M. Horsin Helene, "Technologies of electrical energy storage systems 1 'electric energy." 2019.

[22] Allart David, "Management and electrothermal modelling of Lithium-Ion batteries." These de doctorat, Normandie Universite, 2017.

[23] Celier Cantor; Quentin Courseaux; Flavie Deslandes, "Regulation of the charge of a battery on a solar panel." 2014.

[24] Mohammed Bencherif, "Modelisation des parametres d'une batterie Plomb-Acide, et son intégration dans un systeme photovoltaique autonome" Memoire de master en physique, Universite Abou Bekr Belkaid- Tlemcen 72p, 2015

[25] "Solar Batteries for Energy Storage," 63p.

[26] Dekkiche Abdelillah, "Modele de batterie generique et estimation de I etat de charge." Memoire de master de physique, Universite du Quebec, 2008.

[27] MIT Electric Vehicle Team, "A Guide to Understanding Battery Specifications" 2008.

[28] Ricaud Alain, "Photovoltaic Modules and Systems." 2008.

[29] "Technical HandbookValve-Regulated Lead-Acid Batteries. 19p.

[30] R. D. Issad Idir, Lalouni Sofia, "Modelisation and simulation of a storage system dedicated to photovoltaic power plants." 2014.

Printed by Books on Demand GmbH, Norderstedt / Germany